PERSONAL WIRELESS COMMUNICATIONS

IFIP - The International Federation for Information Processing

IFIP was founded in 1960 under the auspices of UNESCO, following the First World Computer Congress held in Paris the previous year. An umbrella organization for societies working in information processing, IFIP's aim is two-fold: to support information processing within its member countries and to encourage technology transfer to developing nations. As its mission statement clearly states,

IFIP's mission is to be the leading, truly international, apolitical organization which encourages and assists in the development, exploitation and application of information technology for the benefit of all people.

IFIP is a non-profitmaking organization, run almost solely by 2500 volunteers. It operates through a number of technical committees, which organize events and publications. IFIP's events range from an international congress to local seminars, but the most important are:

- The IFIP World Computer Congress, held every second year;
- open conferences;
- working conferences.

The flagship event is the IFIP World Computer Congress, at which both invited and contributed papers are presented. Contributed papers are rigorously refereed and the rejection rate is high.

As with the Congress, participation in the open conferences is open to all and papers may be invited or submitted. Again, submitted papers are stringently refereed.

The working conferences are structured differently. They are usually run by a working group and attendance is small and by invitation only. Their purpose is to create an atmosphere conducive to innovation and development. Refereeing is less rigorous and papers are subjected to extensive group discussion.

Publications arising from IFIP events vary. The papers presented at the IFIP World Computer Congress and at open conferences are published as conference proceedings, while the results of the working conferences are often published as collections of selected and edited papers.

Any national society whose primary activity is in information may apply to become a full member of IFIP, although full membership is restricted to one society per country. Full members are entitled to vote at the annual General Assembly, National societies preferring a less committed involvement may apply for associate or corresponding membership. Associate members enjoy the same benefits as full members, but without voting rights. Corresponding members are not represented in IFIP bodies. Affiliated membership is open to non-national societies, and individual and honorary membership schemes are also offered.

PERSONAL WIRELESS COMMUNICATIONS

IFIP TC6/WG6.8 Working Conference on Personal Wireless Communications (PWC'2000), September 14–15, 2000, Gdańsk, Poland

Edited by

Józef Woźniak
Jerzy Konorski
Technical University of Gdańsk
Poland

KLUWER ACADEMIC PUBLISHERS
BOSTON / DORDRECHT / LONDON

Distributors for North, Central and South America:
Kluwer Academic Publishers
101 Philip Drive
Assinippi Park
Norwell, Massachusetts 02061 USA
Telephone (781) 871-6600
Fax (781) 871-6528
E-Mail <kluwer@wkap.com>

Distributors for all other countries:
Kluwer Academic Publishers Group
Distribution Centre
Post Office Box 322
3300 AH Dordrecht, THE NETHERLANDS
Telephone 31 78 6392 392
Fax 31 78 6546 474
E-Mail <services@wkap.nl>

Electronic Services <http://www.wkap.nl>

Library of Congress Cataloging-in-Publication Data

IFIP TC6/WG6.8 Working Conference on Personal Wireless Communications (2000: Gdansk, Poland)
Personal wireless communications : IFIP TC6/WG6.8 Working Conference on Personal Wireless Communications (PWC'2000), September 14-15, 2000, Gdansk, Poland / edited by Józef Wozniak, Jerzy Konorski.
p. cm. — (International Federation for Information Processing ; 51)
Includes bibliographical references and index.
ISBN 978-1-4757-1020-5 ISBN 978-0-387-35526-9 (eBook)
DOI 10.1007/978-0-387-35526-9
1. Personal communication service systems—Congresses. I. Wozniak, Józef. II. Konorski, Jerzy, 1952– III. Title. IV. International Federation for Information Processing (Series) ; 51.

TK5103.485 .I35 2000
621.3845—dc21 00-058399

Softcover reprint of the hardcover 1st edition 2000

Printed on acid-free paper.

Contents

List of Contributors ix

List of Committee Members xi

List of Reviewers xiii

Preface xv

Wireless Internet Architectures: Selected Issues (*invited paper*)
Adam Wolisz 1

A Modified CDMA/PRMA Medium Access Control Protocol for Voice Users in LEO Systems
Abbas Ibrahim, Samir Tohmé 17

Packet Scheduling in Wireless LANs – A Framework for a Noncooperative Paradigm
Jerzy Konorski 29

MAC Protocol for Wireless ATM – Channel Reservation Methods
Andrzej Stelter 43

Quality of Service Aspects of Transport Technologies for the UMTS Radio Access Network (*invited paper*)
Heba Koraitim, Günter Schäfer, Samir Tohmé 53

Resource Allocation in a Cellular CDMA Environment
R. Bolla, Franco Davoli, Marco Perrando 67

An Improved Speech and Channel Coding for GSM System
Dariusz Godyń, Dominik Rutkowski 79

A Picocellular CDMA/TDD Overlay on GSM
Piotr Kaczorek, Dominik Rutkowski 89

Design of Interoperability Checking Sequences Against WAP
O. Kone, J-P. Thomesse 101

A Comparative Study on Distributed Location Management Strategies in Wireless Networks
Hoang Nguyen-Minh, Harmen R. van As 111

Resource Allocation in Cellular Wireless Systems (*invited paper*)
Villy B. Iversen, Arne J. Glenstrup 123

Evaluation of Traffic Carried by ATM Wireless Access Link Controlled by MEDIAN Protocol
Andrzej Beben, Wojciech Burakowski, Piotr Pyda 133

Minimum GPRS Bandwidth for Acceptable H.261 Video QoS
Iyad Al Khatib, Anders Franzen, Fabio Moioli 147

A Performance Analysis of IEEE 802.11 Networks in the Presence of Hidden Stations
Marek Natkaniec, Andrzej R. Pach 157

An Overview of Activities on Wireless Networks in the European Project COST 257 (*invited paper*)
Wojciech Burakowski, Udo Krieger, Kenij Leibnitz, Andrzej Beben, Michela Meo, Tolga Ors, Jorge Garcia-Vidal, Markus Fiedler 169

End-to-End and Redirection Delays in IP Based Mobility
Jon-Olov Vatn 199

Agent Based Seamless IP Multicast Receiver Handover
Jiang Wu, Gerald Q. Maguire Jr 213

Predistortion for Solid State Amplifier of Mobile Radio Systems
Henryk Gierszal, Witold Hołubowicz, Przemysław Sułek 227

Adaptive Antenna Technique for Mobile Communication
Ryszard J. Katulski 239

Robust Noise Reduction and Echo Cancellation
Kristian Kroschel, Martin Heckmann 249

Estimation of The Channel Impulse Response for GSM System
Jacek Stefański 259

Keyword Index 269

List of Contributors

Al Khatib I., 147
van As H.R., 111
Beben A., 133, 169
Bolla R., 67
Burakowski W., 133, 169
Davoli F., 67
Fiedler M., 169
Franzen A., 147
Garcia-Vidal J., 169
Gierszal H., 227
Glenstrup A.J., 123
Godyń D., 79
Heckmann M., 249
Hoang Nguyen Minh, 111
Hołubowicz W., 227
Ibrahim A., 17
Iversen, V.B., 123
Kaczorek P., 89
Katulski R.J., 239
Kone O., 101
Konorski J., 29
Koraitim H., 53
Krieger U., 169
Kroschel K., 249
Leibnitz K.,
Maguire Jr. G.Q., 213
Meo M., 169
Moioli F., 147
Natkaniec M., 157
Ors T. 169
Pach A.R., 157
Perrando M., 67
Pyda P., 133
Rutkowski D., 79, 89
Schäfer G., 53
Stefański J., 259
Stelter A., 43
Sułek P., 227
Thomesse J.-P., 101
Tohmé S., 17, 53
Vatn J.-O., 199
Wolisz A., 1
Wu J., 21

List of Committee Members

PROGRAM COMMITTEE

Jozef Woźniak, Technical University of Gdańsk (Poland) – Program Chair
Jan Slavik, Testcom Prague (Czech Republic) – IFIP-TC6 WG6.8 Chair
Kiryl Boyanov, Bulgarian Academy of Science (Bulgaria)
Wojciech Burakowski, Warsaw University of Technology (Poland)
Sathish Chandran, Perwira Ericsson (Malaysia)
Imrich Chlamtac, University of Texas at Dallas (USA)
Franco Davoli, University of Genoa (Italy)
Veikko Hara, Telecom Finland (Finland)
Takeshi Hattori, Sophia University (Japan)
Witold Hołubowicz, ITTI Poznań and UTA Bydgoszcz (Poland)
Villy Baek Iversen, Technical University of Denmark (Denmark)
Jerzy Konorski, Technical University of Gdańsk (Poland)
Ulf Körner, Lund University (Sweden)
Demetres Kouvatsos, University of Bradford (UK)
Willie W. Lu, Siemens AG (USA)
Gerald Q. Maguire, Royal Institute of Technology (Sweden)
Olli Martikainen, Helsinki University of Technology (Finland)
Andrzej Pach, Cracow University of Mining and Metallurgy (Poland)
Sergio Palazzo, University of Catania (Italy)
Guy Pujolle, University of Versailles (France)
Dominik Rutkowski, Technical University of Gdańsk (Poland)
Debashis Saha, Jadavpur University (India)
Tadao Saito, University of Tokyo (Japan)
Wojciech Sobczak, Technical University of Gdańsk (Poland)

Otto Spaniol, RWTH Aachen (Germany)
Samir Tohmé, ENST Paris (France)
Ioannis Viniotis, North Carolina State University (USA)
Adam Wolisz, Technical University of Berlin (Germany)

ORGANIZING COMMITTEE

Tomasz Janczak, Technical University of Gdańsk (Poland)
Jerzy Konorski, Technical University of Gdańsk (Poland)
Wojciech Molisz, Technical University of Gdańsk (Poland)
Krzysztof Nowicki, Technical University of Gdańsk (Poland)
Wojciech Sobczak, Technical University of Gdańsk (Poland)
Jozef Woźniak, Technical University of Gdańsk (Poland)

List of Reviewers

Kiryl Boyanov
Wojciech Burakowski
Sathish Chandran
Andrzej Czyżewski
Franco Davoli
Witold Hołubowicz
Villy Baek Iversen
Jerzy Konorski
Ulf Körner
Demetres Kouvatsos
Willie W. Lu
Gerald Q. Maguire
Andrzej Pach
Sergio Palazzo
Guy Pujolle
Mirosław Rojewski
Dominik Rutkowski
Roman Rykaczewski
Debashis Saha
Tadao Saito
Wojciech Sobczak
Otto Spaniol
Samir Tohmé
Ioannis Viniotis
Adam Wolisz
Jozef Woźniak

Preface

There are numerous factors contributing to the dynamic growth of wireless communication systems we've been observing in the past 10 years, the most important being the increasing network user mobility and the technological advances in high-speed data transmission over radio channels. Research centres and standards-making institutions the world over conduct works on 3G integrated systems of person-to-person and person-to-computer communications, wireless counterparts of classical LAN, ATM and IP architectures, satellite and access networks as well as advanced service platforms like WAP and other concepts.

Among the many commercial and non-profit organisations professionally involved in the development of the new information infrastructure, of particular influence is the International Federation for Information Processing. Within its Technical Committee TC-6, a working group WG 6.8 has been set up to co-ordinate IFIP activities in the area of wireless communications. It has done so, among others, by arranging regular meetings of academic and industrial researchers, known as IFIP TC-6 WG 6.8 Workshops on Personal Wireless Communications (PWC). Such workshops were held in recent years in Prague, Frankfurt/M, Tokyo and Copenhagen, and their success has resulted in the promotion of PWC to the status of IFIP Working Conference.

This year's PWC'2000 is hosted by the Faculty of Electronics, Telecommunications and Infomatics of the Technical University of Gdańsk, Poland, and the volume we're now introducing to you contains all the 17 contributions accepted for publication, out of 23 submitted, along with 4 invited papers by prominent researchers active in the area. These should constitute a solid basis for fruitful discussions and hopefully will stimulate an interesting exchange of views on the evolution of wireless systems in the years to come.

We believe that the relaxed end-of-summer atmosphere of the picturesque millennium-old Gdańsk, the home of Hevelius, Fahrenheit and Schopenhauer, will be one more thing to remember PWC'2000 for.

It is our obligation and pleasure to express gratitude to the IFIP bodies sponsoring the Conference, TC-6 and WG 6.8 and their Chairmen, for encouragement and advice. Much credit should go to the Program Committee members and other reviewers who, through their detailed screening of the submitted papers, helped shape up the final program. Finally, our thanks extend to the supporting organisations and the authorities of the Technical University of Gdańsk.

Józef Woźniak
Jerzy Konorski

Wireless Internet Architectures: Selected Issues[1]

Adam Wolisz
Technical University of Berlin, Telecommunication Networks Group

Key words: Wireless, Internet, Mobility

Abstract: After discussion of both: basic issues related to wireless data transmission and internet principles we identify and discuss fundamental problems of their merge. Following this discussion a vision of a specific approach and architecture for organizing wireless internet access called AMICA is outlined.

1. INTRODUCTION

Usage of internet services not only becomes a habit, in fact both in their professional and everyday life people become increasingly dependent on the access to these services. Or might achieve significant profit – both in the business and quality of life sense – if such access would be possible frequently enough. In fact, there are good reasons to consider moving all the communication solutions to the internet platform. Wireless transmission technologies have definitely a potential to contribute to the deployment of easy accessible, flexible internet access. The merger of wireless transmission with the already established internet paradigm appears, however, to be more complex than it might have been expected at the first glance. Therefore this merger is recently one of the hottest research topics in the area of telecommunication networks. In order to focus our discussion let us start

[1] This work has been supported in part by the German Ministry of Research (BMBF) within the Project IBMS in the research area ATM Mobile, as well as by grants from German Telecom, DFG within the Graduate College "Communication Based Systems" and ICSI, Berkeley. More details can be found under http://www-tkn.ee.tu-berlin.de.

with an informal, intuitive definition of the notion of wireless internet access.

For the sake of simplicity let us constrain ourselves to a very classical kind of terminals: just laptops or personal digital assistants. Let us also assume, that terminals can use some kind of digital wireless transmission to/from another device, called further an access point, being in turn connected to the world-wide internet using fixed lines. We assume further that the real data exchange takes place in periods called further sessions, separated by idle periods. Initiation of each session will be mostly triggered by terminal itself, but might be also triggered by some other systems, called corresponding systems.

Our further considerations will be structured as follows: In section 2 we will discuss some basic scenarios and identify major general differences between wireless and wired access. In section 3, we will recall basic internet paradigms, consider what are the implication of introducing a wireless hop in internet, and discuss the different possible meanings of the notion of internet access. Finally in section 4 we will outline AMICA - our specific system vision, with special attention being paid to the transport layer services.

2. WIRELESS ACCCESS

Let us first introduce three different scenarios for the usage of wireless access in general:

In the first scenario, which we will refer to as **basic wireless access** the motivation for using wireless technologies is mainly avoiding installation of cable, which might be kind of cumbersome. In this scenario terminals can be moved exclusively within the range of a single, always the same access point, and the movements are slow (if any)[2]. Let us think here for example about a swimming pool area, or just university courtyard. On the other hand, even using the terminal at home (say in ones favorite armchair, or at the kitchen table) gives good reasons for wireless transmission. One could consider this variant as generalization of cordless telephony. The major challenge in this case is assuring the proper quality of service to the terminal. Numerous wireless transmission technologies are already available, or will

[2] The transmission range might be, technology depend, short or large. By the way: wireless access is attractive also in the case of no movement at all- this case is in the classical telephony referred to as the WLL- wireless local loop. Due to fixed positioning of the terminals relative to the access point several techniques for improvement of the quality of signal might be applied. We will not discuss this case in depth in this paper.

soon become available, for the support of such scenario: Wireless LANs, Bluetooth, IrDA belonging to the most frequently mentioned ones.

In the second scenario, which we will refer to as **nomadic wireless access** the terminal is expected to be moved over distances essentially exceeding the transmission range of a single access point. It is assumed, that multiple access points will be deployed over some area (which might be a building, a campus, a city or a continent) creating islands of connectivity around these access points. We assume for this scenario that terminals may switch between the access points only between the consecutive sessions. This movement takes usually times which are long as compared to the session duration (one might think here in the terms of a scientist visiting in turn several universities). In fact there are no hints in which location – close to which access point – the terminal might appear after movement[3].

Let us stress that multiple parameters like supported bit-rate, error rate, the maximum speed of mobility and the supported range around the given access point are in general not identical even among individual access points supporting the same technology (due to static or dynamic set-up differences). Assuring a simple set-up in the new environment seems to be a major challenge for this scenario (different access points in distant locations might even support heterogeneous technologies). An additional challenge is assuring reachability under the original address in the actual, temporary environment as well as security considerations. This problem is frequently referred to as roaming.

Finally in the third scenario, which we will refer to as **true mobile access**, dynamic changes of the supporting access point during a session, usually referred to as handover, are expected to appear (possibly even several times during a single session). For mobile access to be attractive, the deployment of the access points should be more dense as for the nomadic wireless access, usually a significant (although not necessarily all) part of the area will be in the range of at least one of these access points. We frequently use the notion of coverage while referring to the ratio of the area being within the range at least one access point to the whole area under consideration. In this scenario.

The grade of service continuity in spite of handover is one of the essential quality features for this scenario. Continuity of service might be expressed in terms of no information loss during the handover, sometimes even so called seamless handover, i.e. handover not observable by the user at

[3] It should be pointed out that the principle of nomadic access is in general NOT necessarily to be discussed in the context of wireless communication, nomadic computing can be also considered using wired transmission. It seems, however, that wireless transmission might encourage broader deployment of nomadic computing- think for example in the terms of passengers in airports or scientific conference participants

all, is required. This requirement might result in a necessity for the terminal to remain during the handover in the range of both participating access points. In any case the requirement for frequent, possibly interruption-less handover implies usually a homogeneous system concept in which all the access points (and the end-system) are incorporated. In addition it seems almost natural (although not necessary!) to assume that also the transmission techniques will be always unified. In fact this is the case for the majority of solutions deployed or considered today, like GSM, GPRS or the emerging UMTS.

WIRELESS ACCESS DESIGN ISSUES

During the discussion of wireless access scenarios, we have referred several times to the range/coverage issues. In fact it might seem that the general design goal might be developing of technologies supporting possibly high bit/rates within a possibly large range. We will now discuss reasons why this is not the case:

Any wireless communication uses a given band of frequencies which – according to basic communication theory rules – has to be proportional to the targeted bit-rate. This band is shared with any other device (not necessarily communication device!) which can emit frequencies belonging to this band within the same area[4]. With some simplification one might think of a frequency spectrum in a circle defined by the transmission range in the terms of a single precious resource, which might be shared, but not multiplied. A notion of system capacity, expressed for a given frequency band in (Mbits/s)/(Mhz*square_mile) is frequently used. Assuming a constant range, there is an obvious trade in usage of a frequency band between the number of users and the bit-rate available for each of them.

Increasing the capacity – for example to support a higher number of users with given bit-rate and quality within a fixed range around an access point is achieved usually by acquiring additional frequency bands. This is not simple from the regulatory point of view, and in any case expensive (see the recent press news about frequency auctions for UMTS). This is an essential difference as compared to the case of fixed networks, where capacity can be increased in an (almost) unlimited way just by deployment of additional cables (mostly LWL).

Obviously one can alternatively achieve an increase of wireless system capacity by reducing the radius of connectivity. Note however, that assuring constant coverage in spite of radius reduction leads to dramatic increase of the number of required access points, in addition supporting true mobile

[4] The usage of any band of frequencies has either to be licensed for a given type of equipment or even operator, or subject to specific rules of spectrum sharing in unlicensed mode (so called ISM bands)

access unavoidably increases the handover frequency. Another option for increasing capacity is using some kind of spatial discrimination (that means communication in only some directions), with the progress in smart antennas research [21] this approach should become increasingly attractive.

Finally, the aspect of size and energy consumption is usually important in design of mobile terminals. In fact both the energy consumption during active transmission as well as energy consumption while remaining just ready for receiving have to be strongly limited. The former can be achieved by decreasing the range of communication (which is somehow in line with the phenomena discussed above) or – within acceptable limits – by decreasing QoS (mainly bit-rate and error rate as resulting from lower signal-to-noise ratio). The later one can be achieved by putting the end system temporarily in a „sleeping" mode during which the device is not available for communication.

Last, but not least, the clear trend to support communication with huge number of small devices causes strong pressure for low cost solutions.

Because of the essential differences in the bit-rate/range/power/QoS combinations supported by different technologies on one hand, and the strong push for high economy of spectrum usage (see the recent frequency spectrum auctions!) on the other hand, the diversity of deployed technologies will remain quite impressive. Thus we assume, that a single terminal will have to be able to use alternatively one out of multiple different technologies. In fact as one of the first steps in this directions mobile phones, able to switch between DECT cordless and GSM cellular have already being constructed. Technically this might be achieved by using multiple radio interfaces, however the more attractive option is offered by following the soft-radio concept [22].

3. FUNDAMENTAL ISSUES OF WIRELESS INTERNET ACCESS

First of all we have to decide what we really understand under the term internet access, a concept being recently used in several different ways, and discuss what makes the wireless access/mobile access so different.

Let us first recall the real fundamental concept of internet, following US Federal Networking Resolution from October 24 1995 [13].

"Internet" refers to the global information system that:

(i) is logically linked together by a globally unique address space based on the Internet Protocol (IP) or its subsequent extensions/follow-ons;

(ii) is able to support communications using the Transmission Control Protocol/Internet Protocol (TCP/IP) suite or its subsequent extensions/follow-ons, and/or other IP-compatible protocols

(iii) provides, uses or makes accessible, either publicly or privately, high level services layered on the communications and related infrastructure described herein

Let us stress in context of this definition a couple of issues important for our further discussion.

ad(i) IP Addresses have a hierarchical structure: their bind the end-systems (hosts) to a cluster called class A, B, C network. This feature is very useful for hierarchical routing: the global routing is concerned only with identifying the route to the proper network, establishing the route to the individually addressed hosts within the proper network (or- more frequently to some smaller clusters – sub-networks, like a classical ethernet) is a local issue. This does, however, imply that the hosts can (without additional measures) be only reached by IP packets if it remains within the proper network.

ad(ii) The Internet concept supports a model of non-controlled resource sharing. At the IP level, being the common denominator, there is only one, very limited approach to overload avoidance: just dropping packets. Which might cause their retransmission, or loss of larger application units. It is frequently overlooked, that the basis for stable operation of the internet is the TCP congestion control. This works fine, as long as TCP traffic remains the prevailing part of the total internet traffic. Recently a lot of research efforts aim at developing solutions for enforcing a similar behavior from UDP based applications (see [16]). In fact even an idea of introducing for each end-system a kind of universal „congestion manager" controlling the total amount of data transmitted by this end-system into the internet has recently been presented [2].

ad(iii) The fundamental Internet Protocol- IP addresses only the whole end-system, in fact we would mostly like to address individual processes. This granularity, jointly with additional error recognition is given by UDP (for the delay sensitive error non-sensitive applications), while granularity jointly with reliable delivery is supported by TCP (for error sensitive, delay non-sensitive traffic): the two transport layer protocols operating on top of IP. In effect essentially all the applications operate on top of TCP/UDP rather than IP and know nothing about the protocol operation.

Applications do not see protocols, they see a service interface. In the case of internet, these are predominantly sockets; applications use just UDP and TCP sockets. It is commonly accepted fact, that establishing and widely deploying the socket interface contributed essentially to the growth of amount and diversity of internet applications.

CONNECTING THE END SYSTEM VIA A WIRELESS HOP

As the internet model intentionally avoids any assumptions about how the IP packets will be transported and IP packets are – by definition – forwarded on a best-effort principles, there are no design rules which possible level of link quality might be acceptable or non-acceptable. Thus it is assumed that IP packets should be transportable over any kind of medium. So why should we care especially about usage of wireless communication technologies for accessing the internet? What makes wireless so different?

Due to the nature of the wireless channels, their are essentially more error prone than their classical wired counterparts. Not only is the mean bit error rate generally higher, but the error rate is fluctuating. The most straightforward looking remedy of just introducing a reliable link layer protocol with ARQ error recovery can only shift the problem, as ARQ translates bursty packet losses [24] into significant additional variable delay.

In the wired links packet losses caused by transmission errors are very rare. Therefore it was only reasonable that in course of the internet development end-to-end packet losses (as well as delay variations!) became a synonym to congestion in the nodes (routers). In fact the already mentioned TCP congestion control, crucial for healthy operation of the internet, uses late /missing acknowledgement discovered by time-outs as the congestion indication.

So we can conclude that the adverse influence of a wireless link in terms of packet losses and/or delay variability dominates the characteristic of the end-to-end packet transfer. As the result TCP will keep executing its congestion avoidance functionality in a „pessimistic“ way: the stability will be assured but with a (possibly strong) decrease of the transmission speed as seen by the application (see [4] for extensive discussion of this effect). Which is very bad in case where the bit-rates are anyway lower in comparison to the wired counterparts. In the context of the earlier comments on required TCP –friendliness of UDP traffic, this pertains also to all "well behaving" UDP traffic.

But not only the loss of packets during transmission creates a potential problem for the internet operation. Handover causes in general delays/interruptions of the packet flow, which might again influence TCP operation [6].

Talking about handover we have to stress also another fundamental issue. As mentioned in comments to (iii) earlier in this section, the hierarchical routing based on hierarchical nature of IP addresses implies that with change of a location and joining another network (in the internet addressing sense!) a new, topologically correct IP address has to be additionally assigned to the mobile host. And – in order to support the availability of this host under the old address, special measures have to be activated, for example following the mobile IP pattern (See [25] for discussion of these issues).

WHAT DOES INTERNET ACCESS REALLY MEAN?

Traditionally, providing internet access could be interpreted as requirement for the configuration of the terminal as the internet host with whole TCP/IP protocol stack and thus with ability to exchange IP packets with the access point. We have already pointed out essential performance problems appearing in this variant, which we will call the **Internet Endsystem Access**. Extensive research activities are focused on development of TCP modifications [14], however convincing proof of their efficiency is still to be provided. In addition the idea of modifying TCP according to features of the physical link is not really in line with the internet philosophy. Alternatively solutions with cross-layer information exchange (loss indication [23] or booster type support [3] are considered.

But these traditional approaches are not the only ones! The internet definition recalled at the beginning of this section opens a wide spectrum of possible interpretation of the notion of internet access. In fact the key statement "makes accessible higher level services" is of key importance.

In fact end-users are usually interested only in high level services: either the standard ones– mostly WWW access and e-mail – or even some services developed especially for a dedicated class of users (like for example specific e-commerce systems) rather than direct utilization of the communication services. This variant will be referred to as **Internet Service Access**.

A good example of a solution following this way of reasoning is the recently deployed WAP protocol stack [15]. The application software installed in the end-system is not able to contact directly WWW-servers with the classical http protocol, but has to use a special translation server, instead. This allows to use between the end-system and the translation server a set of „lean" protocols, in case of WAP optimized for usage on links with low bit-rates, and thus well suited for today GSM wireless access. Major disadvantage of such architecture is a break of the „uniformity": not only different protocols have to be used, but also applications have to be developed separately. Assuming that a WAP terminal could have a higher speed wireless connectivity besides of the GSM (say WLAN connectivity), it still would not be able to communicate directly with www-servers.

It is out of the scope of this paper to discuss extensively other possible variants of interpreting the notion of wireless internet access in the scope of the basic internet definition. One peculiar approach which we consider the most advantageous: the Remote Sockets approach will be shortly discussed within the AMICA concept.

4. AMICA SYSTEM VISION

After this initial discussion, it seems to be obvious, that there is a huge multidimensional space of choices in organizing wireless internet access, and behind any reasonable set of solutions there must be a consistent vision of the requirements and system concept. Several interesting approaches can be found in the literature, see for example [12][20][1][19]. We will now introduce our vision- AMICA: Adaptive Mobile Integrated Communication Architecture.

BASIC ACCESS ISSUES

We believe that it is important to organize the basic access which satisfies the following requirements:

- Only the communication layers should be involved in adjustment to the wireless technology. In line with the comment to (iii) in section 3 this means that we insist on keeping the TCP/UDP socket interface, with unchanged service semantic, available for applications running in the terminal.
- As there are essential differences between the features of the wireless hop and the fixed backbone, there should be the possibility to design and implement totally independent mechanisms for error control, flow control and congestion control for both these parts.

In line with this requirements we have developed the Remote Socket Architecture (ReSoA), presented in the case of TCP in *Figure 1* in comparison with the "classical" approach discussed in the previous section.

The basic idea (see [18] for details) is to move the TCP protocol engine from the terminal to the access point, and use on the wireless part a two-layer protocol structure consisting of **the export protocol** (EXP) and the **last hop protocol** (LHP) to make the socket calls available in the terminal. The design of EXP is wireless technology independent, the goal of this protocol is twofold. On one hand the remote execution of socket calls has to be assured. On the other hand this protocol has to be coupled with the TCP protocol engine in a proper way, assuring that the socket call semantic will remain unmodified. We can achieve that by careful distinguishing between the TCP protocol semantics and the socket service semantic, being the refinement of the earlier. Let us explain the difference using the example of the "send" call issued by the corresponding host in the "classical" architecture. In fact, the issuing application will return from this blocking call as soon as data associated with this call will be copied in the buffer of the TCP engine at the corresponding host. This is totally de-coupled from the reliable operation of TCP- at this point in time a proper TCP segment has not even been send! So how can the sending application profit from the luck of TCP acknowledgement (i.e. TCP reliability) in case of difficulties in passing data over the wireless hop? Well a "send" call pertains in TCP always to some already opened, existing connection. If there will be problems in sending some data, either the connection would be aborted, or a requested later graceful connection close will be denied. Both these events would mean that data send over this connection can not be assumed as reliably delivered.

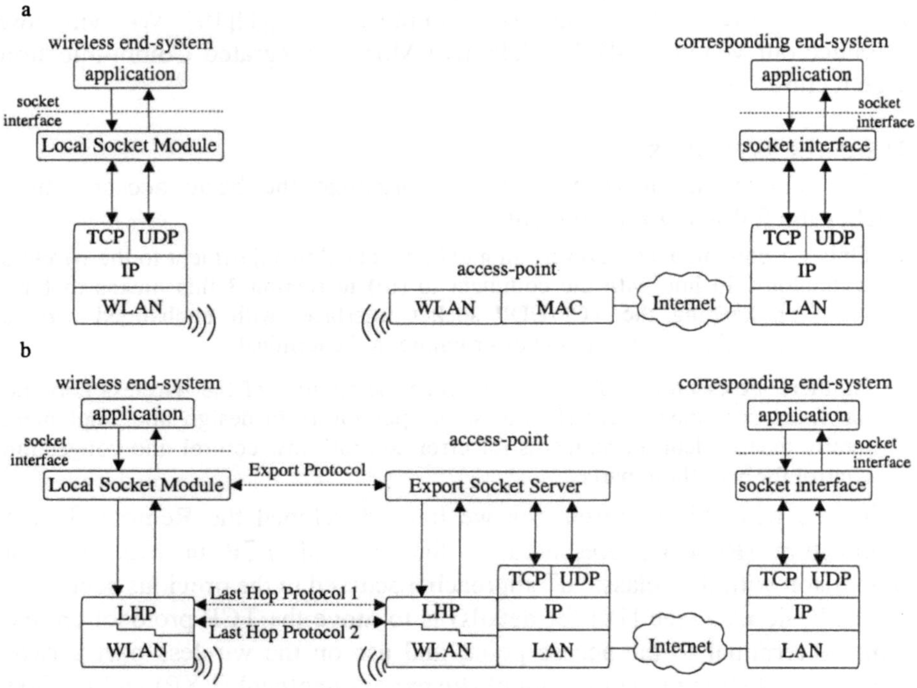

Figure 1. Wireless Internet access using ReSoA

Now, in the case of ReSoA the TCP engine in the corresponding host might get an acknowledgement for the data segment which has been received by the access point, but could not be transferred to the terminal. But the proper coupling of the Export Socket Module of EXP with the TCP engine in the access point will make a graceful close of the TCP connection impossible. Permanent troubles in connectivity between the access point and the terminal might also cause aborting the TCP connection. We believe to have demonstrated with this example the possibility of providing socket service semantic equivalence.

LHP has a different role. This protocol covers the functionality of link layer (including MAC) and is in any case wireless technology dependent. But in addition to that, we believe that spectrum utilization can be maximized, if QoS requested from the wireless link will be adjusted most closely to the real QoS requirements of individual flows. This is difficult in the "classical" architecture, as all IP packets are equal, the situation might improve if QoS supporting architectures, like DiffServ will be agreed upon and supported in wireless links. But in ReSoA we have the possibility to use at least port numbers, and probably additional information available at the transport service interface, in order to differentiate the QoS requirements of individual flows. Different LHP support of TCP flows (reliable LHP) and UDP flows (limited reliability, limited delay) might be a simple example.

There is a huge potential of LHP optimization from different points of view namely QoS support, spectrum efficiency and power saving for the mobile which we follow in a set of research activities. Among others we are investigating optimal use of CDMA type channels (see for example [10][9] for the idea of reducing jitter by using multiple parallel codes for transmission of a single flow), IEEE 802.11 channels (see for example [7] for energy optimization).

Internet end-systems generate usually a strongly asymmetric traffic: majority of the end-systems using high level services get much more data than send, quite a few acting as servers (or sources of data streams – like remote cameras) generate much more data than receive. Knowledge of such traffic patterns can also be mapped on the link layer design. Particularly we expect that exploiting the flexibility of OFDM transmission, being widely accepted for future wireless links might carry a high potential for LHP engineering. Last, but not least, ReSoA has the potential to reduce the protocol processing within the terminal. We will not elaborate this idea.

NOMADIC ACCESS/TRUE MOBILITY: SELECTED ISSUES

Following the considerations from section 3 we believe that the internet access infrastructure will consist of access points supporting very diversified technologies. Most probably, these technologies will differ essentially in the provided coverage, leading to a hierarchy of overlapping cells of different size, see *Figure 2*. Note, the smaller cells will usually not provide complete, 100% coverage of the surface of the overlapping larger cells.

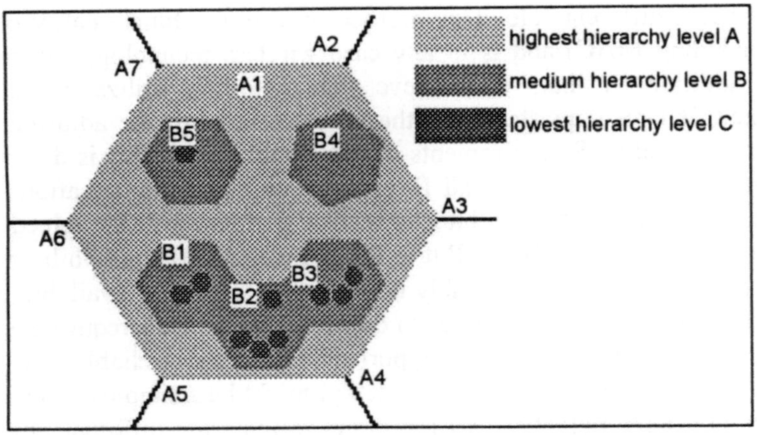

Figure 2. Hierarchical Wireless Network

We assume that access points of all hierarchy levels are connected using high speed fixed links, using Internet protocols (possibly in their emerging QoS enhanced version). However we do NOT assume the existence of a single backbone, like today's public Internet. Contrary, we believe there will be multiple internet backbones with different observed QoS. Access points (or border routers of subnetworks connecting a group of access points) will be able to make decisions as to which backbone to use. This decision will be made on per flow basis, e.g. IP telephony flows might be routed differently from WWW flows.

In order to support multiple communication technologies a mobile terminal has to be equipped – at least conceptually – with a set of different wireless interfaces. In fact as for recent experiments, this is really done using several different physical interfaces (and several device drivers), which definitely is inefficient in terms of size, cost as well as flexibility.

In AMICA we assume that terminals will be soft radio-equipped, and that it will be possible to configure the radio interface according to the actual needs. In addition we assume that within each of the cell type there will be a possibility to select on the fly one out of a set of supported QoS profiles, including bit rate, error rate, spatial coverage and mobility. Examples are easy to point out: in IEEE 802.11 LANs both modulation (equivalent to bit-rate) and power level are adjustable. Furthermore a new technological approach – the smart antennas – seem to give additional boost to such considerations. We will refer to such changes in parameters of communication between a single pair: "terminal- access point" as to **internal handover**. In addition we will talk about **horizontal handover** among access points of the same hierarchy level, and **vertical handover** between access points of different hierarchy levels.

These types of handover differ essentially. In the case of internal handover the access point does not change. In the case of horizontal handover, both the previous and the new access point will (mostly) belong to the same service provider, in the Internet sense they will belong (mostly) to the same network and domain. In the case of the vertical handover, we will usually have to consider a change of the network (in the Internet sense), and even domain: Think in terms of changing from the local network of one of the university institutes, to the GPRS network of some telco. As for the quality of handover mechanisms, we believe that avoiding loss of data and extensive delay should be required. We are, however not necessarily convinced that the existing TCP connections should always be supported, contrary establishing of new connections after handover might frequently be the better choice, especially if full slow start might be avoided. This is subject to vivid investigations.

In AMICA we assume, that for the true mobile access, a terminal is always **at least** in the range of an access point of the highest hierarchical level, the one with the largest coverage, probably most expensive, possibly having lowest QoS (think simply in the terms of GSM-like connectivity). A possibility[5] of vertical handover to a lower hierarchy level (like wireless LAN) will be considered "network initiated" or at least "network supported" as soon as it becomes feasible. We intend to investigate different policies for such handover.

On the other hand, vertical handover from the lower to higher hierarchy level is simple in the sense, that location of a higher level access point "covering" the lower level cell is usually known. But also here a spectrum of different policies as for the decision about handover is to be considered.

To complete the discussion of our philosophy in supporting handover, we would like to stress, that we intend to use extensively the mentioned earlier asymmetry of flows as seen by end systems. As the majority of mobile systems will rather download data, we believe that multicasting data addressed to a target end system several access points which "potentially" might "take over" this target system might essentially reduce the QoS disturbances caused by the handover. Our intention is here to use not only the "classical internet style" multicast of today internet, but also consider some possible modifications (see e.g.[8] for one of our investigations).

[5] One, very attractive way to asses the possibility of handover is the usage of terminal location information (being available from several technologies – see the 911 support as well as newest results in wireless LAN supported location like [5]), in connection with the information on location of lower-level base stations. This seems to be possible not only from GPRS to WLAN but even for handover from, say WLAN to Bluetooth interface...

Contrary to the majority of wireless internet access systems, which follow the "classical" internet philosophy of keeping all the state in the end-system, we believe that it might be advantageous to keep some state information about the mobile systems within the network. In fact, we assume [11] that there will exist a special (logically centralized/physically distributed) repository keeping temporarily, on a soft-state basis, information about all terminals. This information should be used (among others) for supporting handover itself, but also supporting "lightweight" re-authentication after handover.

In fact we believe, that there is a need for a special "logical signaling channel" for supporting the information exchange between the end-system and this repository. This "logical signaling channel" might be mapped at the same hierarchy level access as the data transmission. On the other hand, there are good reasons to consider mapping this logical channel on a connectivity with a higher hierarchical level: the bit rates on this channel will be relatively low, and continuity of it's use in case of either horizontal or vertical handover would be assured. Strategies for assigning this channel to different hierarchy levels will also be considered.

ACKNOWLEDGEMENT

Several previous and recent members of the Telecommunication Networks Group contributed to individual aspects of the AMICA Architecture described here, the contributions of B. Rathke, Morten Schlaeger, J-P Ebert, A. Festag, F. Fitzek deserving a special acknowledgement.

REFERENCES

[1] E. Brewer, et al.: "A Network Architecture for Heterogeneous Mobile Computing", IEEE Personal Communications Magazine, Oct. 1998

[2] H. Balakrishnan et al. "An Integrated Congestion Managment Architecture for Internet Hosts", Proc. ACM SIGCOMM'99, Cambridge, Mass, Sept.1999

[3] H. Balakrishnan, et.al.: "Improving TCP/IP Performance over Wireless Networks", Proceedings of Mobicom, Nov. 1995

[4] H. Balakrishnan, V. Padmanabhan, S. Seshan, R.H. Katz: "A Comparison of Mechanisms for Improving TCP Performance over Wireless Links", IEEE/ACM Transactions on Networking, December 1997

[5] P. Bahl and V. N. Padmanabhan "RADAR: An In-Building RF-Based User Location and Tracking System" Proceedings of IEEE INFOCOM 2000, Tel-Aviv, Israel (March 2000)

[6] R Caceres, V. Padmanabhan "Fast and Scalable Wireless Handoffs in Support of Mobile Internet Radio", Baltzer Journals, November 1997,

[7] J-P. Ebert, A.Wolisz "Combined Tuning of RF Power and Medium Access Control for WLANs", Proc. Of MoMuC'99, San Diego, CA, November 1999

[8] A. Festag, T. Assimakopoulos, L. Westerhoff, A. Wolisz "Rerouting for Handover in Mobile Networks with Connection-Oriented Backbones: An Experimenatal Testbed". accepted for ICATM'2000, June 2000.

[9] Frank Fitzek, Adam Wolisz "Extended Simultaneous MAC Packet Transmission in a CDMA Environment for Quality of Service Support" 3rd European Personal Mobile Communications Conference (EPMCC'99), pp 393-398, March 9-11 1999, Paris, France

[10] Frank Fitzek., Adam Wolisz "QoS Support in Wireless Networks Using Simultaneous MAC Packet Transmission (SMPT)" Advanced Simulation Technologies Conference (ASTC 1999) Featuring Applied Telecommunication Symposium (ATS), pp 185-190, April 11- 15, 1999, San Diego, USA

[11] G. P. Fettweis, K. Iversen, M. Bronzel, H. Schubert, V. Aue, D. Mämpel, J. Voigt, A. Wolisz, G. Walf, J.-P. Ebert "A Closed Solution for an Integrated Broadband Mobile System (IBMS)" International Conference on Universal Personal Communications (ICUPC'96), September/October 1996, Cambridge, Massachusetts, pp 707-711

[12] R.H. Frenkiel, T. Imielinski, "Infostations: The Joy of Many-time, Many-where Communications," WINLAB Technical Report # TR-119, Rutgers University, NJ

[13] http://home.earthlink.net/~serotonin/HotTopicWireless.htm

[14] http://www.ietf.org/html.charters/pilc-charter.html

[15] http://www.wapforum.com

[16] D. Sisalem, A.Wolisz "TCP-Friendly Adaptation: A Comparison and Measurement Study", Proc. NOSSDAV 2000, Chapel Hill, NC, June 2000

[17] M. Schläger, A. Willig: "Improving Wireless Internet Access combining an Active Network Approach and a Proxy Architecture", Technical Report TKN-99-003, Telecommunication Network Group, TU-Berlin, May 1999

[18] M. Schläger, B.Rathke, S. Bodenstein, A. Wolisz; "Advocating a Remote Socket Architecture for Internet Access using Wireless LANs" accepted for publication in Mobile Networks and Applications, Balzer Science Publishers Special Issue on Wireless Internet and Intranet Access

[19] M. Stemm, et al.: "Vertical Handoffs in Wireless Overlay Networks", ACM Mobile Networking (MONET), Special Issue on Mobile Networking in the Internet, Winter 1998

[20] R. Sanchez et al "RDRN: A rapidly Deployable Radio Network- Implementation and Experience", Proceedings of ICUPC'98, Floerence, Italy, October 1998

[21] Smart Antennas, Special Issue of the IEEE Personal Communications Magazine, Vol. 5 No. 1, Feb. 1998

[22] Software Radio, Special Issue of the IEEE Personal Communications Magazine, Vol. 6 No. 4, Aug. 1999

[23] S.M. West, et al.: "TCP Enhancements for Heterogeneous Networks", Technical Report 9-003, Texas A&M University, April 1997

[24] M. Zorzi, R.R. Rao: "Energy Constrained Error Control for Wireless Channels", IEEE Personal Communications, vol. 4, pp. 27-33, December 1997

[25] X. Zhao, et al. "Flexible Support for Mobility", Proc. of Mobicom'98, Dallas, Texas, October 1998, URL http://mosquitonet.stanford.edu/

A Modified CDMA/PRMA Medium Access Control Protocol for Voice users in LEO Systems

Abbas Ibrahim, Samir Tohmé
Ecole Nationale Supérieure des Télécommunications,
Department: InfRes 46, Rue Barrault, 75013 Paris France
aibrahim@enst.fr *Tel: 01 45 81 75 52* tohme@enst.fr *Tel: 01 45 81 78 61*

Key words: LEO satellite channel, CDMA/PRMA, CAC

Abstract: The goal of this paper is to propose a MAC (Medium Access Control) layer to the LEO (Low Earth Orbiting) satellite channel for voice users in order to use efficiently the radio channel bandwidth. This protocol is based on CDMA/PRMA one and adapted to LEO systems. It uses CDMA (Code Division Multiple Access) technique combined with PRMA (Packet Reservation Multiple Access) protocol. A CAC (Connection Admission Control) function that maximizes the number of accepted users with predefined guaranteed QoS (Quality of Service) is then introduced. Furthermore, users traffic control methods are proposed and studied by simulation. The channel parameters are chosen in order to maximize the resource utilization by the proposed protocol.

1. INTRODUCTION

Within personal communication system (PCS), the network needs techniques able to handle a wide range of services with different rates. Furthermore the network will provide a global communication service. LEO satellites constellation will probably be an important component of such a network. Voice application using packet mode is one of these services considered as variable bit rate service due to voice activity feature.

The CDMA technique has been chosen in the third generation mobile network IMT2000. While it does not appear to be a single multiple access technique that is superior to others in all situations, there are characteristics of spread spectrum waveform that give CDMA certain distinct advantages, especially in mitigating multipath fading of the radio link and interference from other systems [5]. Moreover, in CDMA, integration of circuit-mode and packet-mode traffic requires no special protocol, making the use of packet mode to support voice users easy to realize [6].

Section 2 describes the CDMA technique which, will be used in this paper. The third section describes the system model. Section 4 defines voice application at MAC layer. The fifth section presents the channel model. Section 6 presents the proposed access protocol and the CAC function. The traffic control methods are presented in section 7 and compared in section 8. Conclusions are presented in section 9.

2. THE CDMA TECHNIQUE

In this work, the CDMA/DS (Code Division Multiple Access / Direct Sequence) is used with different frequencies in the neighboring cells (spot beams) [10]. A direct sequence code with a very high rate will be added to the original signal as a signature of the transmitter. The receiver will be able to decode this signature and understand the message. By using different frequency bands in neighboring cells, inter cell interference problem [13] is resolved.

In [11], it is assumed that the performance of a CDMA system is dominated by the bit error ratio (BER) performance and problems related to packet acquisition are ignored. A widely used approximation to determine the BER performance on the CDMA channel is the standard Gaussian approximation (SGA) [9]. Assuming that the MAI (Multiple Access Interference) is Gaussian and using simple correlation receivers, the BER or probability of bit error P_e can be obtained from

$$P_e = Q(\overline{SNR})$$

Where: $Q(x) = \frac{1}{\sqrt{2\pi}} \int_x^{\infty} e^{-u^2/2} du$

We consider random direct sequences ($\Pr\{x_j = 1\} = \Pr\{x_j = -1\} = 0.5$) where x_j is a chip of direct sequence with an arbitrary odd codelength (or spreading factor) n. The average signal to noise ratio (SNR) for the ith packet in the case of unequal power reception can be written as

$$\overline{SNR} = \sqrt{\frac{P_i}{(3n)^{-1}\sum_{\substack{k=1\\k\neq i}}^{K} P_k + \frac{N_0}{2T}}}$$

A system with K simultaneous transmitters is considered with received power levels P_j where (j = 1, 2, ..,k), data bit duration T and two-sided spectral density of additive white Gaussian noise $N_0/2$.

In our system no intercell interference will exist because different frequencies are used in neighboring cells. Supposing a perfect power control in the cell, the signal emitted by every transmitter is received by the satellite with P_0 power level, if we neglect N_0 [13]

$$\overline{SNR} = \sqrt{\frac{P_0}{\frac{(K-1)P_0}{3n}}} = \sqrt{\frac{3n}{K-1}}.$$

Assuming that packets with length L bits are transmitted over a memory-less binary symmetric communication channel with average probability of data bit success ($Q_e = 1 - P_e$) and employing a block code, which can correct up to t errors, the packet success probability Q_E can be derived from

$$Q_E = \sum_{i=0}^{t} C_L^i (1-Q_e)^i (Q_e)^{L-i}$$

And by defining a minor limit of the probability of success we can deduce the maximum number of simultaneous users that can use the channel [7] [2].

3. SYSTEM MODEL

The constellation is Iridium like architecture but with different organization of resources. Multibeam antennas are used and each satellite footprint contains a number of spot beams named cells. In each cell uplink and downlink transmissions use two separated radio frequencies, i. e. FDD mode transmission is used. As mentioned in previous section, different frequencies are used in different cells. Assuming that the NCC (Network Control Center) is in one of the gateways. One important function of the NCC is to provide the CAC. The NCC is supposed to have a global view of the network resources. Furthermore, congestion and contention problems are resolved individually at each satellite by using a local control.

4. MAC DEFINITION OF VOICE APPLICATION

At high layer we talk about services supported by the network, for example voice service. At the MAC layer we should talk about capabilities [8]. The mapping between services and capabilities must be defined. In general, service categories are classified to be either real time or non real time. For example in ATM context there are two real time services, CBR (Constant Bit Rate) and VBR-rt (Variable Bit Rate-real time) and three non real time services VBR-nrt, ABR (Available Bit Rate) and UBR (Unspecified Bit Rate). In DiffServ defined with IP (Internet Protocol) there are three differentiated services, premium service for real time applications, assured service and best effort service for non real time applications.

Voice service with voice activity detection can be considered as a real time variable bit rate and can be supported by VBRrt in ATM context and by premium service in IP context. In this article ATM is used and voice service is supported by VBRrt capability and uses AAL2 [12].

5. CHANNEL MODEL

The transmission time scale is organized in frames. Each contains a fixed number of slots. The frame rate is identical to the rate of active voice packets. All transmitters transmit their packets such that they arrive at the satellite within the slot boundaries.

The channel is composed of a succession of frames. Each frame contains a number of slots m. Each slot can support a maximum number of codes n which are used simultaneously (*Figure 1*). One special code is reserved for signaling in each slot and does not influence data codes. A code in a slot is named sub slot. The system contains $M = (m \times n)$ sub slots.

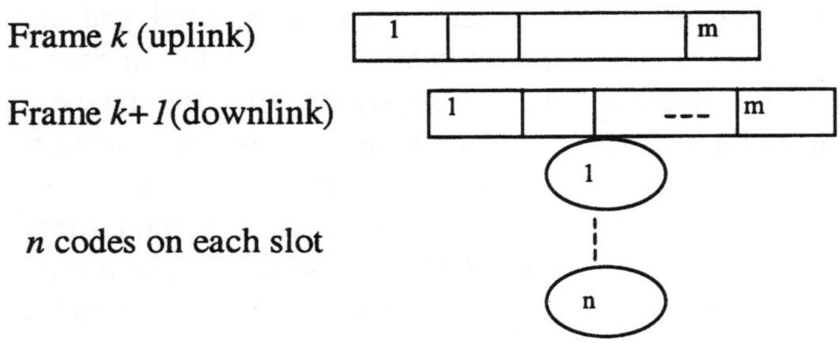

Figure 1. Channel model

In fact the number of codes used in each slot may exceed n and with acceptable quality. This is due to graceful degradation model. According to this model, there is a non-zero probability of correct reception of any arbitrary number of packet [7]. Moreover, the graceful performance degradation enables the addition of users and the mitigation of errors in protocol operation or in network monitoring without the risk of failure of the system operation.

6. CHANNEL ACCESS AND ADMISSION FUNCTION

For voice users there is two important activities: the arrival of a new voice user and the arrival of a voice talkspurt. When a user arrives to the system, he transmits a bandwidth demand request packet. After successful reception of this request, the satellite forwards the request to the NCC and responds to the terminal.

In case the NCC has accepted the demand, the terminal a has then to contend on slot i to transmit his first packet. He does, first, a Bernoulli experiment with P_{ia} probability calculated in relation with the information broadcast by satellite, then he decides to send or not. He switches from contention to reservation mode as soon as he realizes that he sent on a convenient slot. At the beginning of each talkspurt the terminal repeats the process.

Notice that some packets may be lost at the beginning of each talkspurt. They are added to the set of dropped packets. This set is composed of dropped packets at the beginning of talkspurts and erroneous packets due to CDMA interference.

In contrast to PRMA defined in [3] and PRMA-HS (Hindering State) presented in [4], terminals do not classify slots as either "reserved or available" because the channel access for contending terminals is governed by time-varying permission probability. Another difference with PRMA is the CAC function. In our case a direct and simple one is used. A request for setting up a new voice connection will be accepted if there are less than $M \times f$ accepted conversations in the cell, where M is the number of sub slots in the channel and f is a factor higher than one and depends on the access protocol used.

7. THE TRAFFIC CONTROL

In order to fully utilize the channel capacity while maintaining the quality of voice service acceptable, probability of allowing transmission in each slot i for such a user a, P_{ia}, needs to be dynamically updated according to the information about the system state available within the satellite and broadcast to users.

In contrast with [1], P_{ia} is calculated in the mobile station and not in the satellite, that makes it more flexible and simplifies the satellite design. In fact, P_{ia} is set according to the number of voice users who used the slot i in the previous frame. The purpose of this function is to control the total number of users in every slot, such that, the throughput is maximized without exceeding the loss limit, which means, without exceeding the total number of codes allowed to be used simultaneously.

Assume that the satellite, by using its local memory, knows the number of simultaneous codes used in each slot in the last frame MI_i. We have then to choose the function which calculates the probability of sending on a slot i, this function depends on MI_i and should minimize the probability of collision on the slot, that means should distribute users on slots as uniform as possible. Two methods to calculate these probabilities are presented.

1. The first one is the following. Taking into account that each user knows the number of codes used on each slot in the previous frame, user calculates permission probability by:

If $MI_i < n$ then $P_{ia} = p - tg(\alpha) \times MI_i$,

Otherwise $P_{ia} = max(p - tg(\alpha + \beta) \times MI_i, 0)$.

Where n is the spreading factor, tg() is the tangent function, p is the initial probability, α and β are the slopes of the two linear segments represented by the two functions above. These parameters should be chosen to maximize the throughput.

In [1], authors have studied an efficient protocol which uses this computation but with different strategy. In this protocol, user has to be acknowledged to switch to reservation mode. If user is not acknowledged he must retransmit. He can retransmit until the waiting time exceeds "D_{max} = *20ms*" which is the threshold of waiting time for voice user. For satellite communication, this is impossible because of the high round trip time which is comparable with D_{max}. Two variations are introduced.

The first is at the protocol scale: instead of waiting the acknowledgement to switch to reservation mode, the user waits the slot state broadcast by satellite. If the number of codes used on this slot is less than the spreading factor, the user continues the usage of the same slot, if not he must re-contend.

The second is for values of p, β and α,. In [1] the values chosen are "$p = 0.3$, $\alpha = 0.007$ and $\beta = 0.1$". These values are bad in LEO context and produces high dropping probability at the beginning of each talkspurt. The reason is that in LEO system the packet can not wait more than one frame time and the user tries to send this packet on slots dedicated for voice in only one frame time. The new choice is then "$p = 0.8$, $\alpha = 0.02$ and $\beta = 0.2$".

2. The second method is inspired from the fact that, in CDMA used, the maximum number of simultaneous codes which can be used with acceptable quality of service (loss < 0.01) exceeds the spreading factor. So we define two thresholds of control: the spreading factor n and the maximum number of accepted codes in a slot s. The method is as follow:

If the value of n is larger than MI_i, a voice user, in order to begin a talkspurt, uses a code in the slot i with probability $P_{ia} = 1$.

If $n \leq MI_i \leq s$, then $P_{ia} = n/(MI_i \times U)$

Where U is a constant higher than one and must be chosen to have the maximum efficiency of the protocol.

In other cases $P_{ia} = 0$.

After sending the first packet, the user switches to reservation mode if the number of codes used on this slot is less than s. In the other case he must re-contend.

8. SIMULATION RESULTS

These simulations compare different methods presented in section 7 with different parameters of channel. The simulator tool (NS Network Simulator) is used to define CDMA/PRMA channel supporting the proposed protocol. The LEO system used is Iridium like architecture with multibeam technology. In each beam we have 24 frequency carriers and each one has a downlink bandwidth equal to uplink bandwidth equal to 512Kb/s. Gateway (GW) access bandwidth is 155,52Mb/s. ATM cells are used to support voice packets and therefore data field is 48 bytes length and correction field is 10 bits length. Voice rate in the "on" state is 8Kb and channel parameters are assigned two different values:

In the first case, the following parameters are used

- number of slots = 8
- spreading factor = n =512Kb/8Kb/8 = 8
- accepted codes = s = 10 (from computation in section 2 with $Q_E \geq 0.99$)

The second case considers the following values:

- number of slots = 4
- spreading factor = n =512Kb/8Kb/4 = 16

- accepted codes = s = 20 (from computation in section 2 with $Q_E \geq 0.99$)

The mean time of the state ON (T_{ON}) is set to one second and the one of OFF state (T_{OFF}) is fixed at 1.35 seconds. The simulation run time is such that each user is in conversation during 3mn, which is equal to the mean time of voice connection. That necessitates many hours of simulation time and makes the simulation confidential.

Figure 2 compares the two control methods for the two choices of channel parameters. This comparison leads to determine the maximum number of acceptable users in each case. This number determines the limit used by CAC function. As this number augments as the efficiency of the protocol increases:

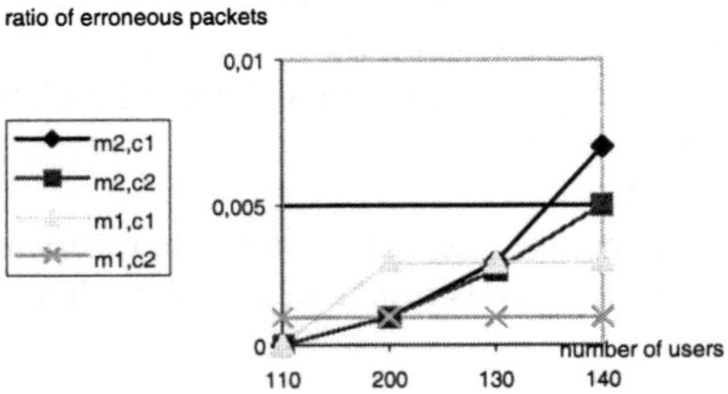

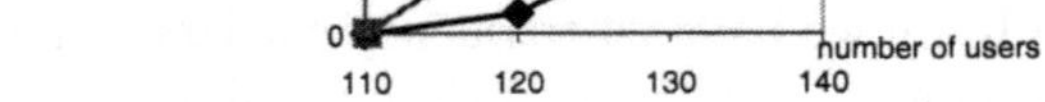

Figure 2. Comparison of methods and choices (mi, cj = method "i" for choice "j")

The second graph in *Figure 2* is the significant one and proves that the second method with first choice gives the best results. With the first method the first choice is better. In the first graph we see the ratio of erroneous packets due to CDMA interference only. In this graph we notice a lower error ratio in the first method because the limit of accepted users on slots have not been attained in the first method due to a high drop probability. Anyway, the main criterion of performance is the total loss ratio. To clarify this discussion *Figures 4* and *Figures 5* present distribution of users in different cases.

Figure 4 shows that the first method is acceptable and the limit of users in a slot (10 users) is attained. On the contrary, the limit (20 users) is not attained in the second choice because of a high drop probability.

Figure 5 illustrates that the second method does significantly improve the utilization of resources and the limit of users on slots is frequently attained and especially for the first choice. This improvement allows augmenting the number of accepted users up to 135 for the first choice as presented in figure 2. This number represents the threshold of CAC function. This threshold varies from 110 users to 135 users and f defined in section 6 varies from 1.7 to 2.1 and its value depends on the control method.

An important performance issue is the multiplexing efficiency relative to perfect statistical multiplexing. We define the multiplexing efficiency factor as: $\mu = M_{0.01} \times \delta / m \times n.$

Where $M_{0.01}$ is the number of simultaneous conversations supported with a loss probability less than 0.01. δ is the voice activity factor given by $\delta = T_{ON} / (T_{ON} + T_{OFF})$ and (m, n) are the slot number and the spreading factor respectively. Table 1 lists this factor for different cases as well as for PRMA and PRMA-HS protocols. This table shows the enhancement of multiplexing efficiency for the proposed protocol with a good choice of the channel parameters.

Table 1: multiplexing efficiency comparison

Protocol, choice, method	**Multiplexing efficiency factor (packets/slot)**
CDMA/PRMA, 1, 1	0,73
CDMA/PRMA, 2, 1	0,75
CDMA/PRMA, 1, 2	0,9
CDMA/PRMA, 2 , 2	0,87
Classical PRMA	0,67
PRMA-HS	0,73

Finally, *Figure 3* presents the effect of channel parameters on the efficiency of the protocol. These parameters can be represented by the spreading factor. When the spreading factor varies between 64 and 1, the system varies from pure CDMA to pure PRMA

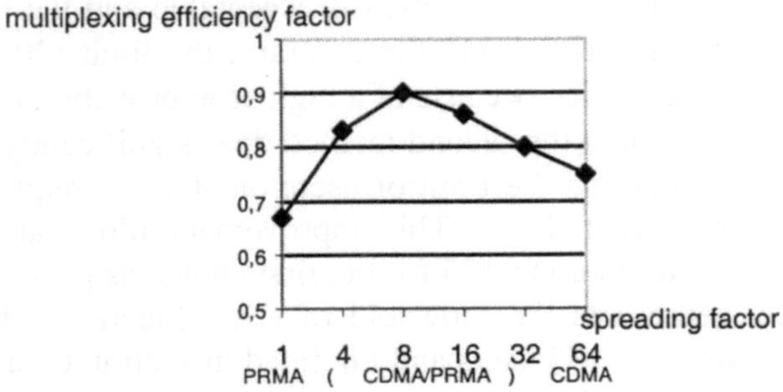

Figure 3. comparison of different channel parameters (second method)

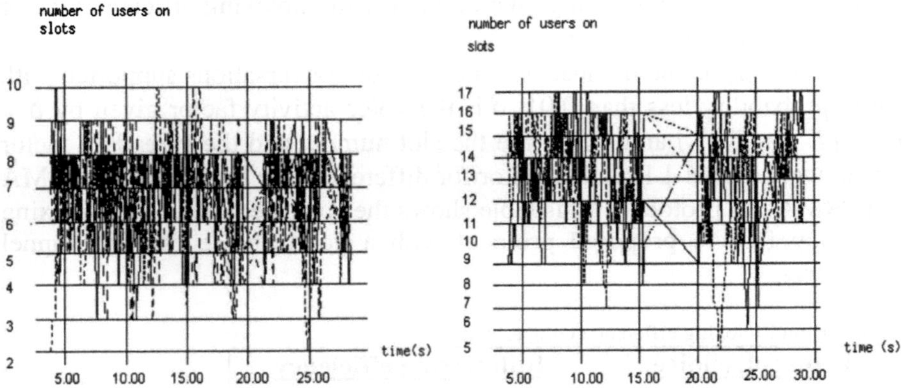

Figure 5: Distribution of 135 users on slots (first and second choices, second method)

9. CONCLUSION

In this paper, a modified CDMA/PRMA protocol is proposed in order to provide access to radio channels in LEO systems. Control algorithms which calculate dynamically the permission probability are then proposed. Choices of channel parameters are presented in order to choose the best one. The comparison of the control methods with various choices gives numerical results which, find out the best method and the best choice. These results illustrate how very high multiplexing efficiency can be achieved by using two thresholds control algorithm with a spreading factor equal to 8. This choice improves the proposed protocol in term of the efficiency of using resources and in term of the increase of system capacity.

REFERENCES

[1] A. E. Brand and A. Hamid Aghvami "Performance of Joint CDMA/PRMA Protocol for Mixed Voice Data Transmission for Third Generation Mobile Communication" IEEE J. on Select. Areas. in Comm. December 1996.

[2] C. LI and R. D.Gitin "Multicode CDMA Wireless Personal Communications Networks" IEEE ICC'95.

[3] D. J. Goodman, R.A. Valenzuela, K.T.Gayliard and Ramamursh "Packet Reservation Multiple Access for Local Wireless Communications" IEEE Trans. on Comm. August 1989.

[4] D. Re Enrico, Romano Fantacci, Giovanni Giambene, and Walter Sergio. "Performance Analysis of an Improved PRMA Protocol for Low Earth Orbit-Mobile Satellite Systems" IEEE Trans. on Vehicular Technology, MAY 1999

[5] G. Evaggelos., Yu-Wen CH., Wen-Bin Y. "Optimal Strategies for Admitting Voice and Data Traffic in Networks of LEO Satellites using CDMA" Wireless Networks 2(1996) 315-328.

[6] J. Abbas. "Low Earth Orbital Satellites for Personal Communication Networks". Artech House. Boston, London.

[7] K. S.Gilhousen, I.M. Jacobs, R. Padavoni and L.A. Weaver "Increased Capacity using CDMA for Mobile Satellite Communication". IEEE J. on Select. Areas. in Comm. 8 (4) 1994.

[8] K. Heba and T. Samir "Resource Allocation And Connection Admission Control in Satellite Networks" IEEE J. on Select. Areas in Comm. February 1999.

[9] M. Gerard and M. Bousquet "Satellite Communication System" third edition. WILLEY 1998.

[10] M. B. Pursley, "Performance Evaluation for Phase Coded Spread Spectrum Multiple-Access Communication part-I: system analysis" IEEE Trans. on Comm. August 1977.

[11] N. D. Wilson, R. Ganesh, K. Joseph, and D. Raychaudhuri, "Packet CDMA versus Dynamic TDMA for Multiple Access in an Integrated Voice Data PCN" IEEE J. on Select. Areas in Comm. August 1993.

[12] K. Sriram, Y. terng wang "Voice Performance Using AAL2 and Bit Dropping Performance and Call Admission Control" IEEE J. on Select. Areas in Comm. January 1999

[13] V. Andrew J. "CDMA Principles of Spread Spectrum Communication" Addison-Welsey 1995.

Packet Scheduling in Wireless LANs – A Framework for a Noncooperative Paradigm

Jerzy Konorski
Technical University of Gdansk, Poland

Key words: wireless LAN, EY-NPMA, Random Token, noncooperative stations

Abstract: Contention-based packet scheduling policies incorporated into MAC protocols in wireless networks attempt to schedule one packet transmission per protocol cycle and are optimised to reduce the scheduling penalty while distributing the bandwidth fairly among the network stations. The paper points out the possibility of there being some noncooperative stations that, instead of adhering to a common-goal optimum policy, try to maximise their individual service rates to the detriment of the cooperative stations. A framework for noncooperative scheduling policies is postulated with analogies drawn from the auction paradigm; desirable features include verifiability at various levels, fairness and low performance cost. Upon discussing possible noncooperative station behaviour, four single-cycle noncooperative scheduling policies are proposed as modifications of the well-known EY-NPMA and Random Token policies. Results of a preliminary simulation study are presented to demonstrate that these modifications do prevent or at least discourage noncooperative stations from stealing bandwidth from cooperative ones.

1. INTRODUCTION

Scheduling packet transmissions in wireless LANs has been intellectually challenging mainly because of the lack of 'natural' scheduling facilities present in satellite or wired LANs such as immediate network-wide feedback, physical ordering of stations, collision detection by sensing the channel while transmitting, on-the-fly modification of other stations' packets etc. The challenge is aggravated by user mobility, which makes it difficult to track actions of a particular station, and changing or non-existent station id's, as in ad-hoc networks. A large variety of distributed scheduling policies have been devised to be incorporated in the wireless MAC protocols (see [7], [8] for a survey and details of some representative policies). All of them impose a scheduling penalty arising from contention, polling, reservation, distributed election and similar mechanisms that consume a certain portion of the channel bandwidth in the form of control packet transmission, synchronisation to time slots or various collision stand-offs. Thus apart from purely random access protocols, a generic *protocol cycle* consists of a scheduling phase and a packet transmission phase (the former may itself be composed of some sub-phases). An important distinction can be made between *single-cycle* and *multiple-cycle* policies that span, respectively, only one protocol cycle (as in ETSI's HIPERLAN [2] or IEEE 802.11's CSMA/CA [10]) and many consecutive protocol cycles (as in PRMA [5] and token-passing protocols). Single-cycle policies can be further divided into *contention-based* and *reservation-based*; the former schedule one packet transmission per cycle, whereas the latter usually attempt to schedule more (an example is the CRT protocol [3]).

Hereafter we shall focus on single-cycle, contention-based policies. Their main goals are:

- high bandwidth utilisation (minimum scheduling penalty) and
- fairness (equal service rates as perceived by all network stations).

While the contradictory nature of these goals has been recognised since the introduction of high-speed wired LANs [6], rapid development and deployment of MAC technologies add a new dimension to the fairness problem. Namely, fairness has so far been striven for with the silent assumption that all the network stations will adhere to the designed policy and co-operate for the common goal. However, this goal will probably be ruined if noncooperative stations are present and pursue their own goals (i.e., greedy maximisation of individual service rates). Consider a CSMA/CD station that refuses to back off after a collision; a HIPERLAN station that always generates long elimination bursts; a CRT station that sends multiple reservation packets per cycle while only one is allowed – clearly, such stations have a competitive edge on the other, cooperative stations. Suitably modified MAC interfaces (like bogus parts in the aircraft industry) can be easily manufactured and distributed, if half-legally, on a commercial basis. This calls for a new generation of *noncooperative scheduling policies* that prevent noncooperative stations from stealing the bandwidth from cooperative ones. Note that a similar evolution has been observed in WANs and internets: from globally optimised mechanisms (e.g., routing) to architecting independently administered, noncooperative networks [4], often based on game theory and seeking a Nash equilibrium point. This paper is meant as a stimulus to a discussion along these lines.

The material is presented as follows. In Sec. 2, the network model is described along with a simplified specification of two exemplary scheduling policies: EY-NPMA of ETSI's HIPERLAN and Random Token (RT) first proposed in [1]. Sec. 3 discusses possible noncooperative station behaviour and outlines the framework of noncooperative scheduling policies. In Sec. 4, modifications of EY-NPMA and RT are proposed to improve fairness in the presence of noncooperative stations. Sec. 5 contains the results of a simulation study.

2. NETWORK MODEL AND SCHEDULING POLICIES

The following assumptions and non-assumptions will govern our network model:

A1. The network uses a single-frequency radio channel shared on a time-division basis.

A2. There are N network stations of which NC are noncooperative; the stations never go down or switch off.

A3. All stations hear one another (a single-hop configuration).

A4. The maximum station-to-station propagation delay constitutes a common *time slot*; all stations synchronise to the boundaries of successive time slots.

A5. The transceiver at a station, when in the receive mode, is able to sense the channel at various, policy-specific levels of accuracy, distinguishing idle/busy states, idle/single transmission/collision states or measuring the total carrier power on the channel.

A6. Stations need not reveal their id's, or have any permanent id's at all; noncooperative stations do not present themselves as such (thus N and NC remain unknown to all stations).

A7. The employed cooperative scheduling policy (i.e., adhered to by the cooperative stations) is known to all stations.

As examples of cooperative scheduling policies we take EY-NPMA and RTCA (RT with Collision Avoidance, a slightly improved version of RT) specified below in a simplified version. Both are single-cycle, contention-based policies whose viability has been confirmed by numerous implementations (of the former) and analysis (of the latter).

EY-NPMA: at the beginning of a protocol cycle each station examines the contents of its packet buffer and, if nonempty, considers itself active, enters the *elimination phase* and transmits an *elimination burst* of random length (1..E_{max} time slots). Thereupon, if the channel is sensed idle, the station enters the *yield phase* in the next time slot; if the channel is sensed busy, the station backs off (i.e., defers its packet transmission to a future protocol cycle). In the yield phase, a packet is transmitted after a random *yield delay* (1..Y_{max} time slots) unless the channel is sensed busy before that, indicating a packet transmission by another station participating in the yield phase. *Figure 1* illustrates the protocol cycle (note that the declaration of priorities and resynchronisation issues of the original EY-NPMA have been left out).

RTCA: in the elimination phase, an active station delays its packet transmission for a random *elimination timeout* (1..E_{max} time slots), then transmits the packet unless sensing the channel busy with another station's packet before that. Again, resynchronisation issues are left out. However, to allow direct comparisons with EY-NPMA, we shall improve RT somewhat in the spirit of CSMA/CA; namely, instead of a whole packet, a station transmits a 1-slot pilot packet to discourage those active stations who have chosen longer elimination timeouts. Next, it enters the yield phase similarly as in EY-NPMA, along with the other stations that transmitted their pilot packets in the same time slot. The protocol cycle is illustrated in *Figure 2.*

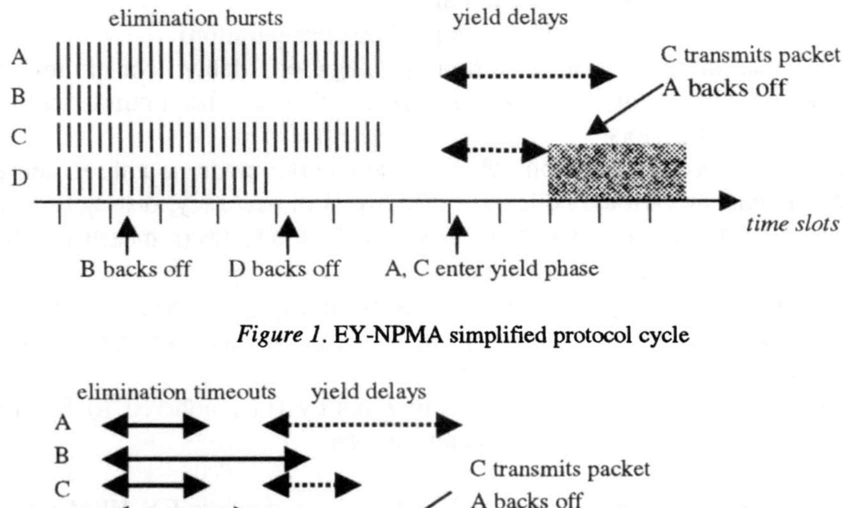

Figure 1. EY-NPMA simplified protocol cycle

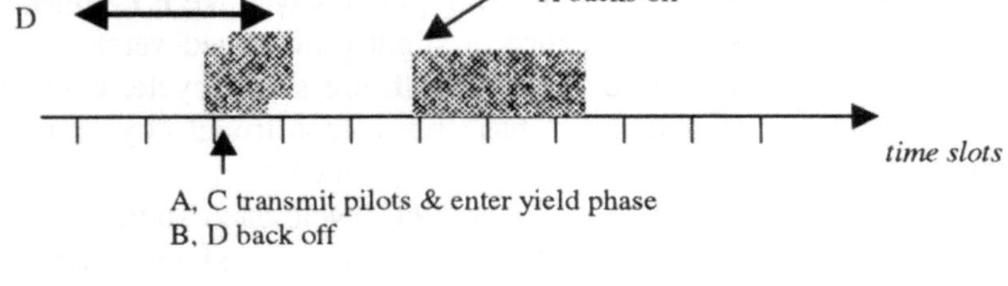

Figure 2. RTCA simplified protocol cycle

It is interesting to view the elimination phase as a kind of an auction where the stations bid their elimination bursts or timeouts. EY-NPMA corresponds to a first-price sealed-bid auction, and RTCA – to a Dutch auction (where the item being auctioned is awarded to the earliest bidder who matches the progressively descending price) [9]. In our scheduling context, these auctions have a few peculiar features:

- the maximum price of the item i.e., the right to transmit a packet, is bounded (by E_{max}),
- some bidders are willing to buy the item at any price and therefore can cheat by choosing their bids deterministically rather than at random, and
- the auctioneer's goal is to make the item equally affordable to all bidders rather than maximise the selling profit.

3. FRAMEWORK FOR NONCOOPERATIVE SCHEDULING

In EY-NPMA, a simple strategy of a greedy noncooperative station could consist in transmitting its elimination burst and subsequently, upon sensing the channel still busy, resuming the elimination burst with a good chance of outbidding the other active stations (which would turn the sealed-bid auction into a classical English one). In RTCA, a noncooperative strategy could consist in joining in the yield phase despite not having transmitted a pilot packet. This raises the issue of *verifiability*.

Suppose a dedicated network station has at its disposal a monitoring device equipped with a small-aperture antenna able to track the activity of any other station it locks upon. It is easy to see that a noncooperative station using any of the above strategies can thus be identified and neutralised by imposing whatever sanctions one might think of (e.g., heavy penalties on the user or just jamming the ensuing packet transmission). A hypothetical noncooperative scheduling policy forbidding the above strategies would then be *small-aperture verifiable*. A stronger verifiability constraint would involve a monitoring device of less channel sensing accuracy e.g., of the idle/busy (carrier detection) type. Now noncooperative stations could use the above strategies and get away with it, making the corresponding noncooperative scheduling policy *carrier unverifiable*. However, another noncooperative strategy could be devised for EY-NPMA whereby a station starts transmitting its packet right after the longest elimination burst has terminated. Similarly, in RTCA a noncooperative station could start transmitting its packet without prior transmission of a pilot packet i.e., not having subjected itself to the elimination phase. A hypothetical noncooperative scheduling policy forbidding these strategies would be *carrier verifiable*. In fact, a taxonomy of noncooperative scheduling policies could be proposed based on the strength of the verifiability constraints they meet. In our study we shall require that a noncooperative scheduling policy be at least small-aperture verifiable. When deciding on its bidding (elimination phase) strategy, a greedy noncooperative station is thus left with small-aperture unverifiable options, given that it is not prepared to risk sanctions imposed by the monitoring device.

One can imagine that a noncooperative strategy will in general consist in bidding longer elimination bursts or shorter elimination timeouts than resulting from the cooperative randomisation mechanism (assume for simplicity that the noncooperative behaviour does not extend to the yield phase). Such a policy can also be soundly assumed to be

– *stationary* – in particular, a noncooperative station will remain so all the time, without ever reforming to become cooperative again,
– *inconspicuous* – attempting to mimic cooperative behaviour in order to conceal its greediness e.g., by randomising bids in successive protocol cycles rather than insistent bidding extreme values in each cycle,
– *rational* – based on the knowledge of the cooperative bidding strategy and aiming to maximise individual service rates rather than to damage the other stations', and
– *isolated* – only aware of a possible presence of other noncooperative stations and not their number, location etc., in particular unable to collude with any other noncooperative station.

In response, a noncooperative scheduling policy should ensure that a noncooperative station does not outperform a cooperative one significantly in the long run, at least for a wide range of *NC/N* ratios. One might allow significant unfairness for small *NC/N* on the premise that the few noncooperative stations will not steal enough bandwidth to worry about and besides, small *NC/N* will not persist too long as more and more cooperative stations, lured by the prospect of stealing some extra bandwidth, turn noncooperative. Likewise, at high *NC/N* unfairness is less harmful since there are few cooperative stations to steal bandwidth from.

Note in passing that detection of noncooperative stations based on monitoring the source statistics of successful packet transmission is out of the question because

– it would span multiple protocol cycles – in fact, a large enough number to maintain statistical credibility,
– inconspicuous noncooperative strategies would make statistical inference even harder, and
– source information is not necessarily deducible from packets (assumption A6 of Sec. 2).

4. NONCOOPERATIVE SCHEDULING POLICIES

Any verifiable noncooperative scheduling policy should prevent forceful bidding (insistent bidding long elimination bursts or short elimination timeouts), at the same time not deterring cooperative stations from bidding at all. Therefore, some straightforward policies have to be ruled out e.g., imposing penalties upon stations whose elimination bursts were longest or exclusion from the yield phase of stations whose elimination timeouts were shortest. Here we propose four modifications: two of EY-NPMA, called EY-NPMA/*(a,b)* and EY-NPMA/2ndMAX, and two of RTCA, called RTCA/1stCOLL and RTCA/1stSINGLE. They differ in the channel sensing accuracy required at each station: idle/busy for EY-NPMA/*(a,b)*, total carrier power for EY-NPMA/2ndMAX and idle/single transmission/collision for the RTCA modifications. See *Figure 3* and *Figure 4* for illustration.

EY-NPMA/*(a,b)*: upon termination of its elimination burst, a station counts the number of time slots when the channel is still sensed busy (with longer bursts). If and only if that number is not greater than a and greater than b, the station is allowed to join in the yield phase (a and b are integer parameters, $E_{max} \geq a > b \geq 0$). Thus only bidders within a predetermined range from the highest bid are the winners.

EY-NPMA/2ndMAX: upon termination of its elimination burst, a station that senses the channel still busy counts the number of power drops around the boundaries of successive time slots (reflecting subsequent termination of longer elimination bursts). If and only if that number is 1, the station joins in the yield phase. Thus only second-highest bids win.

RTCA/1stCOLL: upon transmitting its pilot packet, a station waits for the other stations' reaction in the next slot. If the channel is sensed idle, it joins in the yield phase; otherwise it backs off and all other active stations start the elimination phase anew, with E_{max} decreased by 1. An active station reacts to a single pilot packet by transmitting a 1-slot burst and refrains from reacting if a collision of pilot packets was sensed. Thus in terms of the Dutch auction, an earliest bidder wins only if his bid is not unique.

RTCA/1stSINGLE: similarly as in RTCA/1stCOLL except that active stations react to collisions of pilot packets and refrain from reacting upon single pilot packets. Having transmitted a pilot packet and sensed the channel idle in the next slot, a station starts transmitting its packet immediately. Thus an earliest bidder wins only if his bid is unique.

Commenting from the auction perspective, the above modifications encourage the bidders to outbid each other while ensuring that the item is also affordable to the less wealthy bidders. EY-NPMA/*(a,b)* and EY-NPMA/2ndMAX resemble the 'reverted' Vickrey auction where the second-highest bid wins, but the winner pays the highest bid.

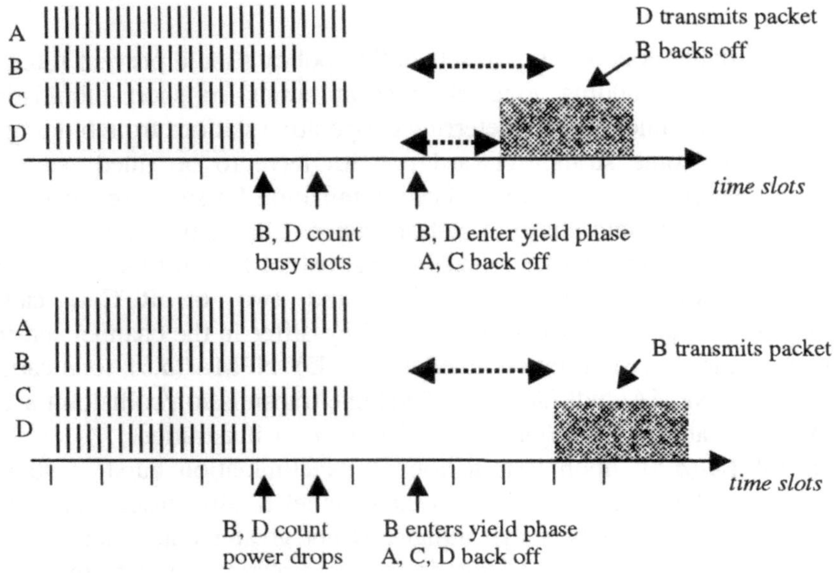

Figure 3. EY-NPMA/(2,0) (*top*) and EY-NPMA/2ndMAX (*bottom*) protocol cycles

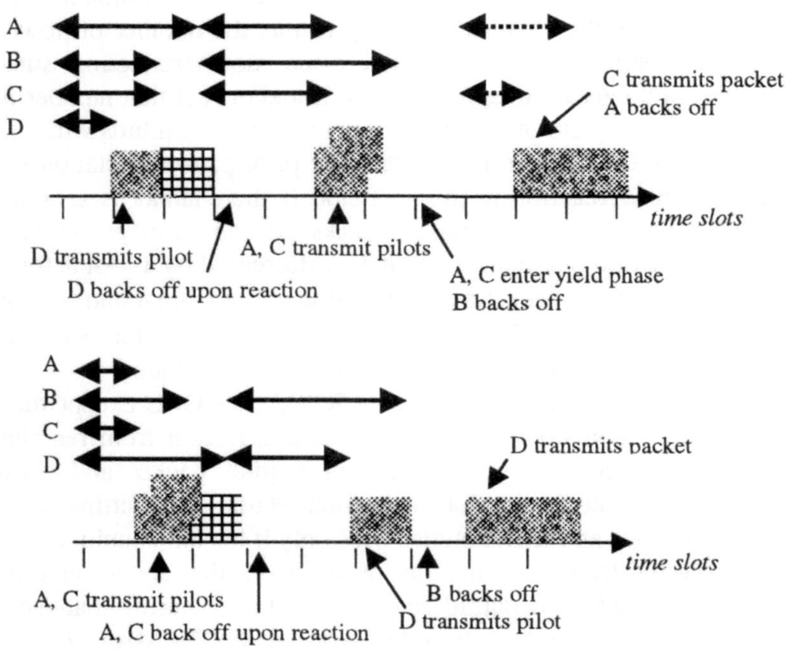

Figure 4. RTCA/1stCOLL (*top*) and RTCA/1stSINGLE (*bottom*) protocol cycles

5. PERFORMANCE STUDY

A simulation experiment has been set up to investigate the performance of the four noncooperative scheduling policies of Sec. 4 and compare them to the basic EY-NPMA and RTCA policies. The network and scheduling policy parameter setting is shown in *Table 1.*

Table 1. Network parameter setting for simulation

number of stations	N=10
number of noncooperative stations	NC=1..10
elimination burst/timeout at cooperative stations	1..E_{max}=15 time slots uniform distribution
elimination burst/timeout at noncooperative stations	1..E_{max}=15 time slots shifted distribution (see below)
yield delay	1..Y_{max}=3 time slots uniform distribution

The noncooperative bidding strategy is assumed to be identical at all noncooperative stations and consist in shifting the distribution of the elimination burst/timeout by an integer constant $1 \le m \le E_{max}$. That is, *elimination_burst*:=$\langle E+m\rangle_{1..E_{max}}$ time slots and *elimination_timeout*:= $E_{max}-\langle E+m\rangle_{1..E_{max}}+1$ time slots, where E is uniformly distributed over 1..E_{max} and $\langle\cdot\rangle_{I..J}$ denotes clipping of a random variable so that it falls into the interval $I..J$.

As a principal performance measure we take the per-cycle average station service rate $P_{succ} = Prob\{$station transmits packet successfully | station active in the present cycle$\}$. Ideally, $P_{succ} = 1/N$, although in practice $P_{succ} < 1/N$ on account of imperfect collision resolution. Note that P_{succ} is indicative, through queuing theory relationships, both of the average packet throughput and packet delay; thus focusing on P_{succ} relieves us from setting specific characteristics of offered load, packet sizes and buffer capacities. However, the choice of E_{max} and Y_{max} is sensitive since one has to weigh the growing efficiency of collision resolution against the degradation of throughput as they increase; the above values have been optimised experimentally.

In *Figure 5* and *Figure 6*, P_{succ} is plotted against NC and m for the basic EY-NPMA and RTCA policies, separately for cooperative and noncooperative stations. As expected, the plots show a high degree of unfairness, with the service rates at cooperative stations quickly dropping to a tiny fraction of the all-cooperative values (at NC = 0) as NC and m increase. On the other hand, noncooperative stations steal a large portion of the bandwidth until they become too many and so have few cooperative stations to steal from.

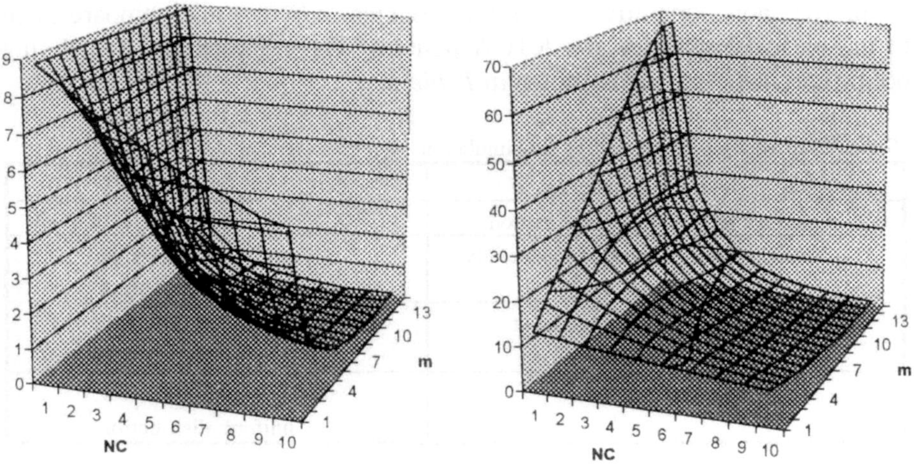

Figure 5. EY-NPMA, P_{succ} in % for cooperative (*left*) and noncooperative (*right*) stations

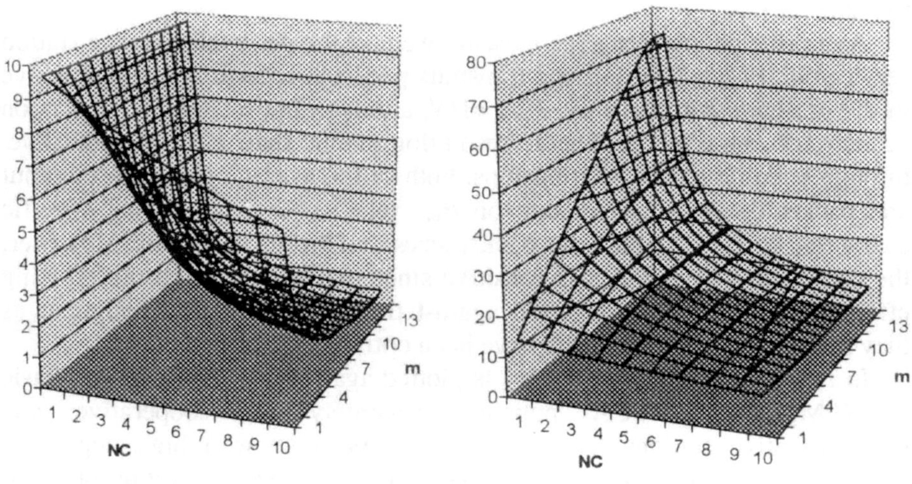

Figure 6. RTCA, P_{succ} in % for cooperative (*left*) and noncooperative (*right*) stations

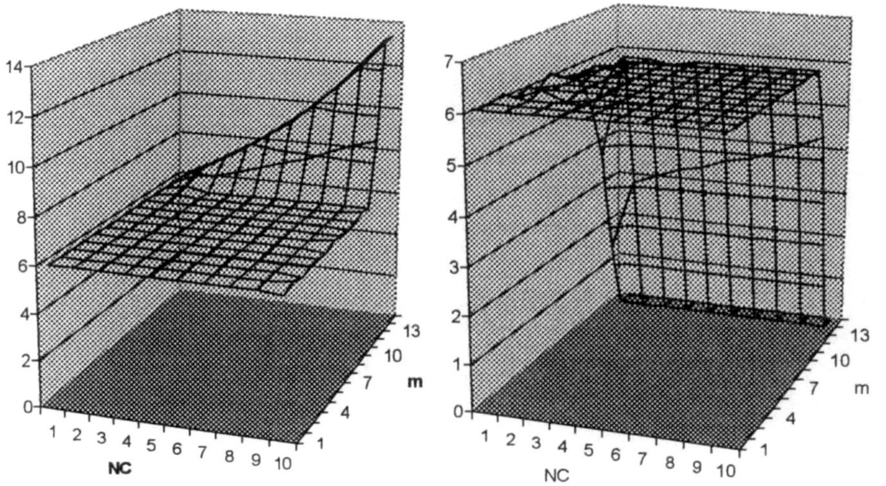

Figure 7. EY-NPMA/(2,0), P_{succ} in % for cooperative (*left*) and noncooperative (*right*) stations

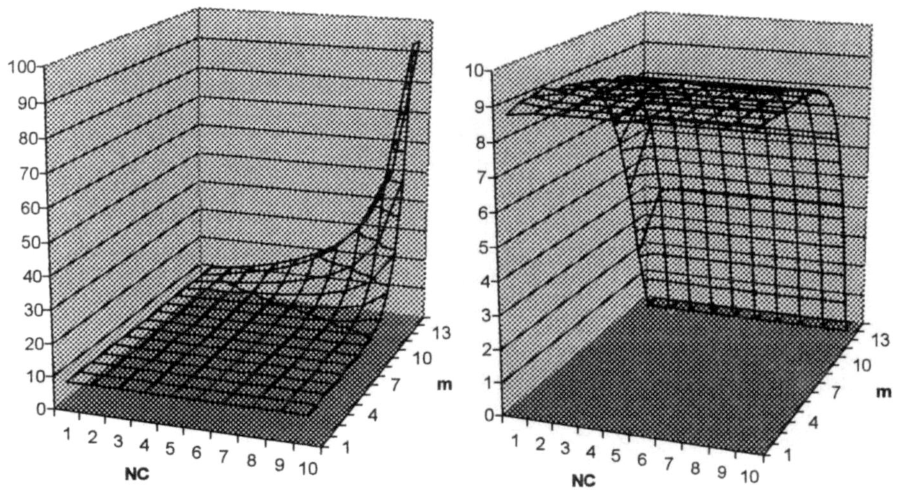

Figure 8. EY-NPMA/2ndMAX, P_{succ} in % for cooperative (*left*) and noncooperative (*right*) stations

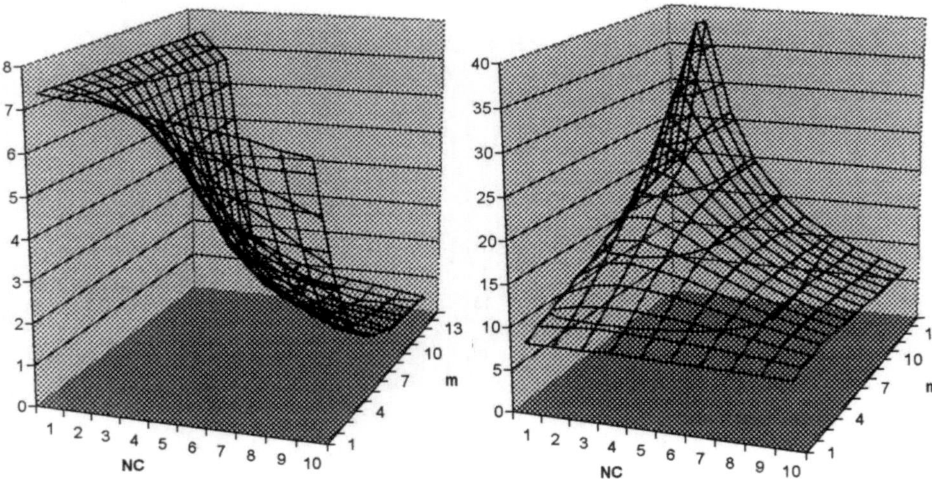

Figure 9. RTCA/1stCOLL, P_{succ} in % for cooperative (*left*) and noncooperative (*right*) stations

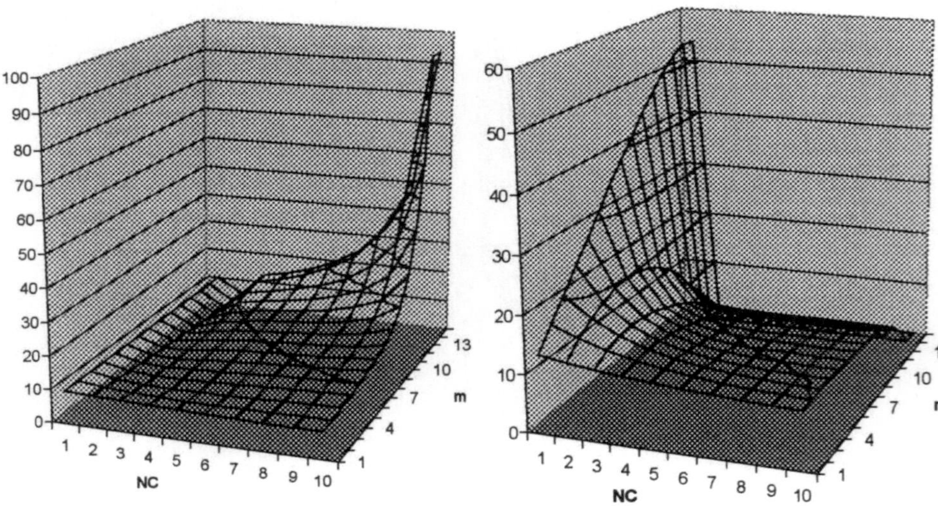

Figure 10. RTCA/1stSINGLE, P_{succ} in % for cooperative (*left*) and noncooperative (*right*) stations

The situation for EY-NPMA is remedied by EY-NPMA/(2,0) (*Figure 7*) and EY-NPMA/2ndMAX (*Figure 8*), which appear an ideal disincentive for a station to become noncooperative: P_{succ} is hardly greater for noncooperative than for cooperative stations; moreover, cooperative stations even gain an advantage when they are few. This comes at a price, however – the all-cooperative P_{succ} is smaller than in *Figure 5* – and the price is higher for EY-NPMA/(2,0) than for EY-NPMA/2ndMAX (a drop from 8.9% to 6.1% vs. 7.9%, respectively). The latter observation stems from the fact that while in EY-NPMA/2ndMAX there are winners of the elimination phase in each protocol cycle, in EY-NPMA/(2,0) there may be none.

RTCA/1stCOLL and RTCA/1stSINGLE (*Figure 9* and *Figure 10*) paint a bit less ideal picture. Although the unfairness is distinctly reduced (cf. *Figure 6*), it still persists at some *NC* and *m* (as another illustration of the difficulty of the Dutch auction design). However, RTCA/1stSINGLE looks quite attractive in that it guarantees reasonable service rates for cooperative stations over almost the whole parameter range, while disfavouring noncooperative stations except for very small (thus not persistent) *NC*/*N*. (Paradoxically, this policy looks more suited for forceful bidding than RTCA/1stCOLL.) The all-cooperative P_{succ} drop (refer again to *Figure 6*) is insignificant for RTCA/1stSINGLE and acceptable for RTCA/1stCOLL – from 9.7% to 7.9%.

6. CONCLUSION

We have pointed out the need for noncooperative scheduling policies in contention-based MAC protocols for single-channel wireless LANs. Like in WAN internets, where the noncooperative management paradigm has already set in, one should account for the presence of noncooperative entities that will pursue their individual goals instead of adhering to a common-goal optimum policy. Building on two well-known protocols, EY-NPMA and RT, and drawing on the auction paradigm, we have proposed four noncooperative scheduling policies and investigated their performance via simulation to confirm that they prevent, or at least discourage, noncooperative stations from stealing the channel bandwidth from cooperative ones. This comes at the cost of a modest drop of the all-cooperative station service rate. Some questions for future research are:

- How can RTCA/1stCOLL and RTCA/1stSINGLE be further improved?
- What would a noncooperative scheduling policy look like in response to a general noncooperative strategy (rather than based specifically on a distribution shift, *m*)?
- Do effective carrier verifiable noncooperative scheduling policies exist?

- How to construct noncooperative counterparts of reservation-based and priority scheduling policies – is there a systematic way of seeking a Nash equilibrium analogue?

REFERENCES

[1] I. Chlamtac and A. Ganz, *Evaluation of the Random Token Protocol for High-Speed and Radio Networks*, IEEE J. Select. Areas Commun., **SAC-5**, July 1987

[2] ETSI TC RES, Radio Equipment and Systems; *High Performance Radio Local Area Network (HIPERLAN); Services and Facilities; Version 1.1*, RES 10, Jan. 1995

[3] J. Konorski, K. Nowicki and J. Wozniak, *The Cooperative Random Token Protocol for High-Speed Radio LANs*, Proc. GLOBECOM'92, Orlando FL, Dec.1992

[4] Y.A. Korilis, A.A. Lazar and A. Orda, *Architecting Noncooperative Networks*, IEEE J. Selected Areas Commun., **13**, Sept. 1995

[5] S. Nanda, D.J. Goodman and U. Timor, *Performance of PRMA: A Packet Voice Protocol for Cellular Systems*, IEEE Trans. Veh. Technol., **40**, Aug. 1991

[6] *Special Issue on Media Access Techniques for High-Speed LANs and MANs*, Computer Networks and ISDN Syst., **26**, March 1994

[7] *Special Issue on Mobile Radio Communications*, IEEE J. Selected Areas Commun., **SAC-2**, July 1984

[8] *Special Issue on Packet Radio Networks*, Proc. IEE, **75**, Jan. 1987

[9] R. Wilson, *Auction Theory*, [in] J. Eatwell, M. Milgate and P. Newman (eds), The New Palgrave, MacMillan, London 1987.

[10] *Wireless Medium Access Control (MAC) and Physical Layer (PHY) Specifications*, Draft Standard IEEE 802.11, P802.11/D1

MAC Protocol for Wireless ATM – Channel Reservation Methods

Andrzej Stelter
Poznań University of Technology, Institute of Electronics & Telecommunications
Email: astelter@et.put.poznan.pl

Key words: Wireless ATM, Medium Access Control

Abstract: The paper describes and evaluates by simulation an efficient channel access protocol (MAC protocol) based on TDMA scheme. Two methods of the buffer status transmission from mobile terminal to base station have been considered and compared. The first method combines a random access channel with piggybacking (R+PGBK), the second one combines a random access channel with backlog collision free slots assigned to the selected connections (R+B). The methods have been compared in two different systems. Simulation results show that for both proposed systems, R+B method is better.

1. INTRODUCTION

In recent years, wireless ATM (WATM) has been a hot topic in telecommunications research. Typically WATM is considered as the wireless sub-domain of the B-ISDN infrastructure. Very important issue in such systems is MAC mechanism. As the RF bandwidth is a very limited resource, an effective channel utilization is extremely important for the wireless systems. In literature there are presented several medium access protocol propositions for WATM [2,3,4]. In majority of cases, ATM transmission is based on TDMA scheme, using TDD for the communication among the base station and the mobile terminals. Multiple access protocol is based on a TDMA time frame. Typically the protocol is centralized, i.e. it is controlled by the base station. The base station assigns the time slots (bandwidth) to the terminals which cooperate with it. To do the job efficiently, the base station must have sufficient knowledge concerning ATM cells waiting in the terminal output buffers. The terminal output buffer status information is transmitted by a short packet in a random access channel. The performance of such protocols depends on several factors [5]:

- uplink channel reservation method (the way in which mobile terminals reserve the uplink channel bandwidth),
- scheduling algorithm (algorithm used by the base station to assign time slots to individual terminals or connections),
- collision resolution algorithm (algorithm performed by terminals whose request packets have collided in a random access channel),
- the length of the frame.

In the paper only the first from the above mentioned four factors is studied. As has been stated, the terminals reserve the uplink channel bandwidth by sending short request packets in a random access channel. Because of possible collisions, such solution does not guarantee that the base station will have the packet on time. This would be particularly harmful for the real time connections. In majority of the WATM multiple access protocols there are implemented additional ways to transmit the buffer status packets in a collision free manner. The particular solutions presented in the literature are difficult to compare. The reason is lack of any common standard concerning simulated system assumptions.

In the simulation presented in this paper there are considered the protocols in which a random access channel is combined with one of two collision free transmission methods, either the output buffer status information is piggybacked onto ATM cell transmitted uplink [3,4], or it is sent in a minislot assigned to the selected connection [2].

2. PROTOCOL DESCRIPTION

In the paper it is considered a wireless network microcell with one base station and several buffered mobiles. There is studied a time-slotted duplex system in which the uplink (mobile-to-base) and downlink (base-to-mobile) communications are time multiplexed on a single frequency channel. The terminals do not communicate directly among themselves. Packets (ATM cells) waiting for transmission in each terminal are stored in the output buffers. Information concerning each buffer state is sent to the base station where it is used to assign a time slot to the connection.

In the considered time slotted system, one ATM cell is transmitted in one time slot. The slots are grouped in frames. The frame structure is shown in *Figure 1*. The frame starts with a header transmitted downlink by the base station. The header contains information about the frame structure and the acknowledgements for packets transmitted in the previous frame. The second field of the frame contains downlink slots D in which ATM cells are transmitted from base station to the mobile terminals. In the next field of the frame there are uplink slots U in which ATM cells are sent from terminals to the base station. Each packet transmitted in U and D slot comprises an ATM cell and two bytes of CRC code. In the frame, besides the uplink (U) and downlink (D) slots, there are short request slots R in which terminals send a buffer status information, i.e. the number of cells in the buffer queue. The R slots create slotted ALOHA type channel. When collision occurs in a slot R, the packet is retransmitted in the next frames according to the binary exponential backoff algorithm (probabilities of the packet retransmission in the subsequent frames diminish in the following way: 1/2, 1/4, 1/8,...).

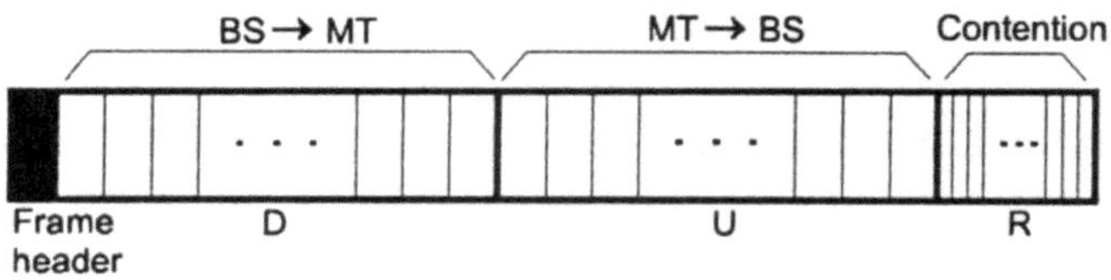

Figure 1. Frame structure

In the presented simulation, terminals are able to send the buffer status information not only in slots R but also

- piggybacking this information (PGBK field) onto transmitted uplink ATM cells (R+PGBK method) or
- sending this information in additional slots B assigned to the rt-VBR connections (R+B method).

Because ATM cells can be buffered in terminals in different ways, the two mentioned above channel reservation methods (R+B and R+PGBK) are compared in two systems, which differ in terminal buffer organization (see *Figure 2*). Because of different buffer organization, the information transmitted in minislots B and R and in the field PGBK is different in both studied systems.

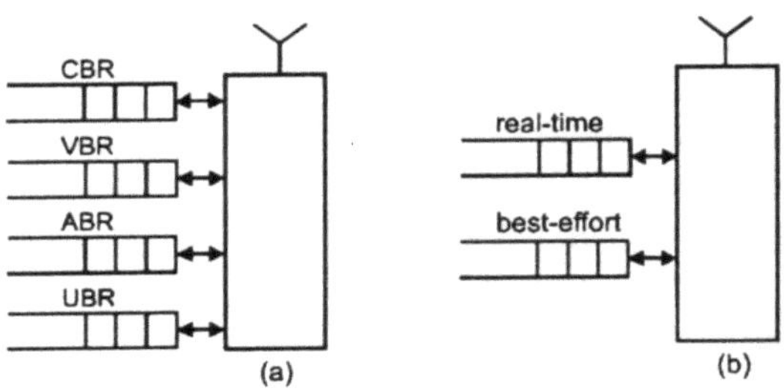

Figure 2. Terminal buffers: a) System I, b) System II

3. SYSTEM I

In System I, each connection has its own identifier unique in the microcell (hence it is not necessary to identify terminals) and each connection has its own buffer in the terminal (see *Figure 2a*). Each ATM cell in a buffer is characterized by a 'delay', which rises with time. Cells in the buffers are arranged in order of delays and always a cell of the maximum delay is the first in a queue. CBR and rt-VBR cells for which the delay exceeds the specified limits are rejected. nrt-VBR, ABR and UBR cells of maximum delay are rejected when their buffers are overloaded.

Two methods of transmission of buffer status information to the base station are studied. In case of R+PGBK (see *Figure 3a*), a slot R is used only when simultaneously two conditions are fulfilled i.e. the buffer assigned to the connection is empty and a new ATM cell is generated. In this method the slot R packet consists of 10-bit connection identifier *connection_id* (it allows 1024 connections per microcell) and 2-bit number *n_cells* which contains information about the number of new cells in the queue (not registered in the base station). Two-bit number piggybacked onto ATM cell contains number of new ATM cells in the buffer.

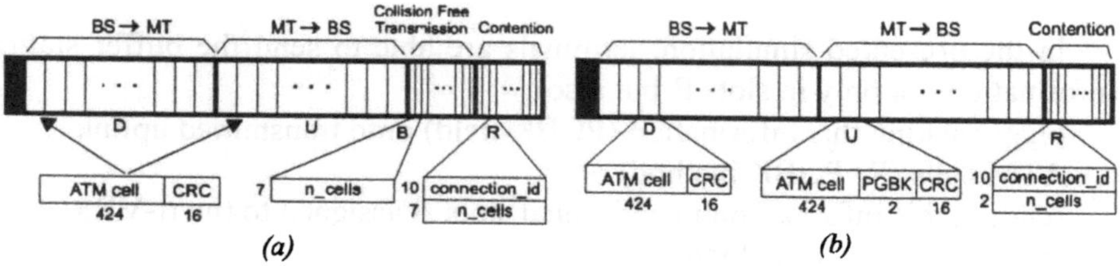

Figure 3. Channel reservation methods (System I): a) R+PGBK, b) R+B

In case of R+B method (see *Figure 3b*), each rt-VBR connection has assigned a minislot B, hence transmission of the buffer status information of the rt-VBR connection is collision free. Each rt-VBR connection has assigned a collision free slot B because rt-VBR traffic is characterized by a large range of changes and strict delay requirements. The seven-bit number *n_cells* in slots R and B is able to assign all frame slots to one connection when it works with maximum speed (it was assumed that the total number of transmission slots D and U in the frame is 100). In case of R+B method the terminals having CBR, nrt-VBR, ABR and UBR connections have to compete for the channel access in slots R.

In the considered centralized system, the uplink slots U in each frame are assigned to the connections by the base station. The base station handles a table, which contains a buffer descriptor for each connection. Each descriptor stores a number of ATM cells waiting for transmission in a buffer assigned to the connection. The table status is brought up to date after each successful packet transmission in the slot R and the buffer status information taken from packet in slot B (R+B method) or the buffer status information piggybacked on ATM packets in slots U (R+PGBK method).

3.1 Simulation description

The simulation results presented in the paper concern the uplink transmission. In the simulation all frames have constant length: 50×D and 50×U slots. It is assumed that the system transmission rate equals 20 Mb/s, therefore one bit is transmitted in 50 ns. It is assumed also that the radius of the wireless network microcell is equal to 45 m, hence the maximum propagation delay equals 150 ns (3-bit interval). Each uplink slot is widened by this interval to avoid overlapping the signals transmitted by different terminals. Under the above assumptions the frame length equals ~2.4 ms. The uplink channel capacity equals ~8 Mb/s.

Two types of terminals that generate three traffic types (CBR, rt-VBR and ABR) are realized in the simulation. The first terminal is a videophone. It transmits both voice and video. The voice traffic represents CBR traffic (during voice activity), the video traffic represents real time VBR traffic. This type of terminal is denoted $T_{CBR+VBR}$. CBR traffic is modeled by a two-state Markov process. It is assumed that a mean length of the voice activity interval equals 1 s, and a mean silence interval equals 1.35 s. In active state CBR source generates a constant bit flow of the rate 32 kb/s. Maximum allowed delay of a cell generated by the above source equals 32 ms. CBR connection informs the base station (in a slot R) only about the first packet in a burst. After successful reception of the request packet, the base station assigns necessary slots periodically, with a constant rate.

rt-VBR traffic source was modeled using the method belonging to the class of discrete–time batch Markov arrival process [1]. For the rt-VBR source it is assumed that a mean bit rate is equal to μ=256 kb/s, standard deviation σ=128 kb/s and autocovariance $C(\tau)=\sigma^2 e^{-a\tau}$ ($a=3.9\ s^{-1}$). Values of the above parameters can describe a video source with no rapid movements in the scene. It is assumed that the maximum delay of ATM cells generated by the above rt-VBR source is 20 ms.

The second terminal type, denoted T_{ABR}, realizes ABR packet data transmission. ABR traffic is modeled by the Poisson process. Each T_{ABR} terminal generates data flow of the rate 256 kb/s. ABR buffer size is 100 KB, hence for the mean rate of 256 kb/s the maximum delay is ~3 s.

In the simulated system CBR connections have the highest priority, ABR connections have the lowest priority. This means that at first the base station assigns slots U to CBR connections, after that slots U are assigned to rt-VBR connections and next slots U are assigned to ABR connections.

All connections from terminals $T_{CBR+VBR}$ and T_{ABR} are established at the beginning of the simulation and are active during the whole simulation process. For a selected input traffic the number of terminals is constant during the simulation. The input traffic in the simulated system is defined by the number of $T_{CBR+VBR}$ and T_{ABR} terminals. The simulation always covers 1200 s of the system activity for a given input traffic.

3.2 Simulation results

The simulation results presented in this chapter were previously published in [6]. The simulation was performed for three different traffic scenarios. The scenarios differ in contribution of CBR, rt-VBR and ABR traffic in the whole input traffic. In all three scenarios the contribution of CBR traffic in the whole input traffic is low. The majority of the input traffic is generated by rt-VBR and ABR connections. In the first scenario the traffic generated by the rt-VBR connections is the same as the traffic generated by the ABR connections. In the second scenario predominates rt-VBR traffic and in the third scenario predominates ABR traffic. The results obtained for all three scenarios are similar. For lack of space only results obtained for the first scenario are presented.

In the first scenario the number of $T_{CBR+VBR}$ terminals is the same as T_{ABR} terminals. The contribution of CBR, rt-VBR and ABR connections in the whole input traffic equals 2.6%, 48.7% and 48.7% respectively. The number of terminals of each type is changed from 8 to 34, hence the input traffic changes from 4.2 to 17.8 Mb/s.

For the R+B protocol a number of slots R equals 7. This number ensures that CBR cell loss ratio is less than 0.01. For R+PGBK protocol a number of the slots R has to increase because they are utilized also by the rt-VBR connections. The simulation results are shown for this number equal 20, 40 and 60. CBR cell loss ratio for the above mentioned numbers is zero.

It is worth to mention that for the whole input traffic range and all protocol versions the average CBR cell delay equals to the length of the frame i.e. ~2.4 ms.

Figure 4 shows the average delay of rt-VBR ATM cells versus the number of terminals. The results show that the delay for R+B method is less than for the R+PGBK method. The reason is that in R+B method the base station is aware of the new rt-VBR ATM cells, waiting for transmission, with the delay at most equal to the length of the frame. In R+PGBK method, rt-VBR connection informs the base station about its buffer status in a slot R. Because of collisions the average delay is greater in this method.

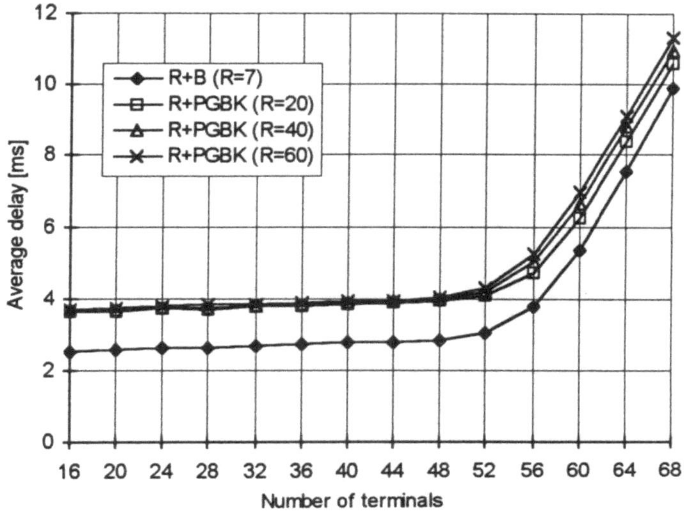

Figure 4. Average delay vs. input load for rt-VBR connections (System I)

The average delay of ABR cells for both versions of the protocol (R+B and R+PGBK) is very similar. It is not presented in the paper.

Figure 5 shows the rt-VBR ATM cell loss ratio. During the simulation, when $T_{CBR+VBR}$ terminals generate input traffic less than the channel capacity, we do not observe any cell rejections for the R+B protocol. A different result concerns R+PGBK protocol. For this protocol, even for small input traffic, we observe some cell rejections. In this case the cell loss ratio depends on the number of the slots R. We can notice that the cell loss ratio starts to decrease when the number of terminals increases and crosses the channel capacity (see *Figure 5*, 32÷44 terminals). This rather strange phenomenon can be explained by the fact that when there are not enough slots U in the system they are assigned mainly to rt-VBR connections because these connections have higher priority than ABR connections. The buffers for ABR connections are filling and ABR connections use slots R very seldom. Hence, even if in the system there are more terminals generating ABR traffic, they use slots R rarely and more of these slots can be used by the rt-VBR connections. With the increase of the input traffic (above 44 terminals) the cell loss ratio increases. From *Figure 5* it can be noticed also that increasing the number of R slots, from 20 to 60, decreases the cell loss ratio. The cost is the greater redundancy in the system.

Figure 6 shows the cell loss ratio for ABR connections. ABR cells are rejected when the number of terminals increases above 28 (above the channel capacity). The lowest cell loss ratio is observed for R+B method because of the smallest redundancy in the system. In R+PGBK method the cell loss ratio increases with the number of the R slots because they diminish the system throughput.

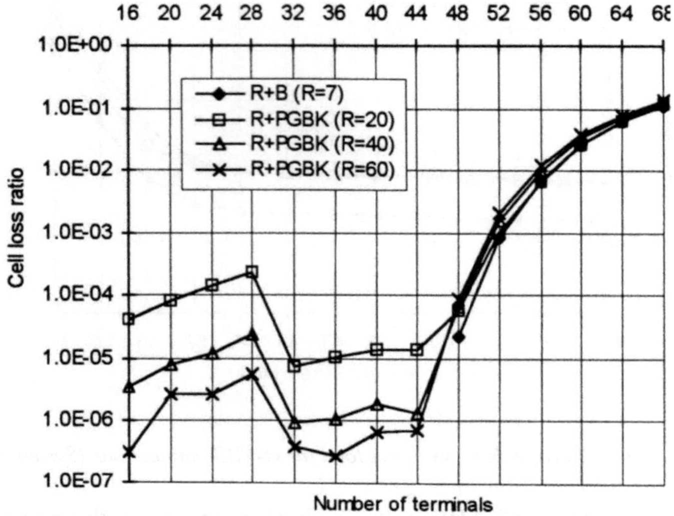

Figure 5. Cell loss ratio vs. input load for rt-VBR connections (System I)

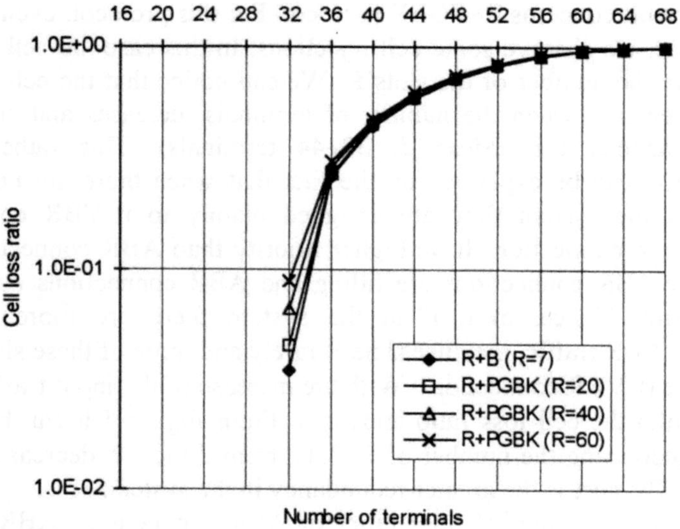

Figure 6. Cell loss ratio vs. input load for ABR connections (System I)

Comparing results shown in *Figure 5* and *Figure 6* for a small traffic (number of terminals less than 32) it can be noticed that the cell loss ratio for ABR connections is better than for rt-VBR connections, even the last ones have higher priority. The reason is that the maximum delay for ABR cells is ~3 s and the maximum delay for rt-VBR cells is 20 ms. Therefore frequent collisions in the channel R force very quickly the terminal realizing rt-VBR connection to reject the oldest cell in the buffer. The terminal realizing ABR connection has more time to inform the base station about its buffer status.

4. SYSTEM II

In System II, the packets (ATM cells) are buffered at the mobile terminals in two queues (*Figure 2b*). In the first queue (*real-time queue*) there are cells from all real-time connections (CBR, rt-VBR). In the second queue (*best-effort queue*) there are cells from other connections (nrt-VBR, ABR, UBR). Cells in both queues are arranged in order of deadlines. Cell of the shortest deadline is the first. There are two similar virtual queues in the base station. These queues keep references to remote packets, which are stored at mobile terminals waiting to be transmitted to the base station. When the base station assigns time slots U to the mobiles, the real-time queue cells have high priority, the best-effort queue cells have low priority. The real-time queue cells are transmitted before the best-effort queue cells.

The same (as in System I) two methods of uplink channel reservation were evaluated by simulation in System II. The simulation parameters were similar to that used in the simulation of System I. The only difference were connection types realized in the system. There were assumed identical traffic characteristics for all mobiles. Each mobile terminal generates rt-VBR traffic and ABR traffic thus in each terminal two kinds of buffers (real-time and best effort) are created. Both rt-VBR and ABR sources were modelled by the methods described in the previous chapter. The simulation was performed for two different traffic scenarios. In the first scenario each terminal generates rt-VBR traffic of the mean rate equal to 256 kb/s (maximum allowable delay τ=10 ms) and ABR traffic of the average bit rate equal to 100 kb/s (τ=500 ms). The number of mobiles varies from 14 to 40, so the system input uplink traffic is equal to ~5÷14.2 Mb/s. In the second scenario each terminal generates rt-VBR traffic of the mean rate equal to 2 Mb/s (τ=50 ms) and ABR traffic of the average bit rate equal to 1 Mb/s (τ=1 s). The number of mobiles varies from 1 to 4, the input traffic rate increases from 3 Mb/s to 12 Mb/s. Results of the simulation of System II were published in [7]. They are similar to that obtained for System I.

5. CONCLUSION

From the results of the simulation we see that for both systems considered in the paper, R+B method to send buffer status information in the uplink channel is better than R+PGBK one. It gives higher system throughput, smaller average access delay and smaller cell loss ratio for all studied connection types. In R+PGBK method the rt-VBR cell loss ratio can be decreased by an increase of the number of R slots, but it makes other parameters of the system (e.g. ABR connection parameters) worse because of higher redundancy in the system.

A disadvantage of R+B method (in comparison with R+PGBK) can be the necessity for terminal having rt-VBR connection to turn on its transmitter one time more in every frame. This can cause larger power consumption.

Conclusions presented above concern a system with the error free wireless channel and perfect synchronization. That is why packets transmitted in B and R minislots are not protected and no synchronization fields are added to them. In practical system however errors will appear, thus packets in B and R minislots should be protected – some redundant bits should be added to these packets. In both methods (R+B and R+PGBK) the number of minislots (B and/or R) does not differ significantly, so it suggests that not taking into consideration redundant bits in the minislots has little influence on the result of comparison of both studied methods. This problem is studied by the author in detail at present.

In practical system all packets should be extended by a header containing a synchronization pattern. That will decrease the system throughput but should not influence the result of comparison of R+B and R+PGBK method.

REFERENCES

[1] C. Blondia, O. Casals, "Performance Analysis of Statistical Multiplexing of VBR Sources", Proceedings of INFOCOM'92, pp. 828-838, 1992

[2] N. R. Figueira, J. Pasquale, "Remote-Queueing Multiple Access (RQMA): Providing Quality of Service for Wireless Communications", Proceedings of INFOCOM'98, pp. 307-314, 1998

[3] N. Passas, S. Paskalis, D. Vali, L. Merakos, "Quality-of-Service-Oriented Medium Access Control for Wireless ATM Networks", IEEE Comm. Magazine, pp. 42-50, Nov. 1997

[4] D. Petras, "Medium Access Control Protocol for Wireless Transparent ATM Access", IEEE Wireless Comm. Systems Symposium, Long Island NY, pp. 79-84, Nov. 1995

[5] A. Stelter, "Medium Access Control Protocols for Wireless ATM System", Poznań University of Technology, PhD thesis, Jan. 2000 (in Polish)

[6] A. Stelter, P. Szulakiewicz, "ATM MAC Protocol for the Radio Interface – Performance Evaluation", Proceedings of PIMRC'99, Osaka, Japan, 1999

[7] A. Stelter, P. Szulakiewicz, "Wireless ATM MAC Protocol – Performance Evaluation by Simulation", Proceedings of WIRELESS 99, pp. 449-455, Calgary, Canada, 1999

Quality of Service Aspects of Transport Technologies for the UMTS Radio Access Network

Heba Koraitim, Günter Schäfer, Samir Tohmé
Ecole Nationale Supérieure des Télécommunications, Paris, France
Email: [Samir.Tohme|Heba.Koraitim|Guenter.Schaefer]@enst.fr

Key words: Quality of Service, UMTS, RAN, AAL-2, IP, Satellite Communications

Abstract: The radio access network of upcoming third generation mobile communication systems is currently a subject of intense research and standardization efforts. This article discusses in the first part quality of service aspects of the realization of the transport network using the currently most concurring technologies for this issue: ATM adaptation layer 2 (AAL-2) and IP. The second part of the paper considers architectural alternatives for realizing the radio access network with satellite communications.

1. INTRODUCTION

The upcoming 3rd generation mobile communication system UMTS is envisaged to be employed in a wide range of scenarios, as depicted in figure 1. The design of its' radio access network (RAN) is currently subject to intense research and standardization efforts. The first approach for the terrestrial radio access network (UTRAN) currently developed by the 3rd Generation Partnership Project uses ATM as a proven technology for transport in the fixed part of the RAN. An alternative, very recent development is to consider an All-IP architecture combined with a quality of service (QoS) framework for this. The specification of a satellite RAN is quite open up to now. This paper presents the current state concerning transport technologies to be deployed in the fixed part of the terrestrial RAN and considers different architectues for a satellite RAN. Special attention is given to the QoS aspects in the RAN.

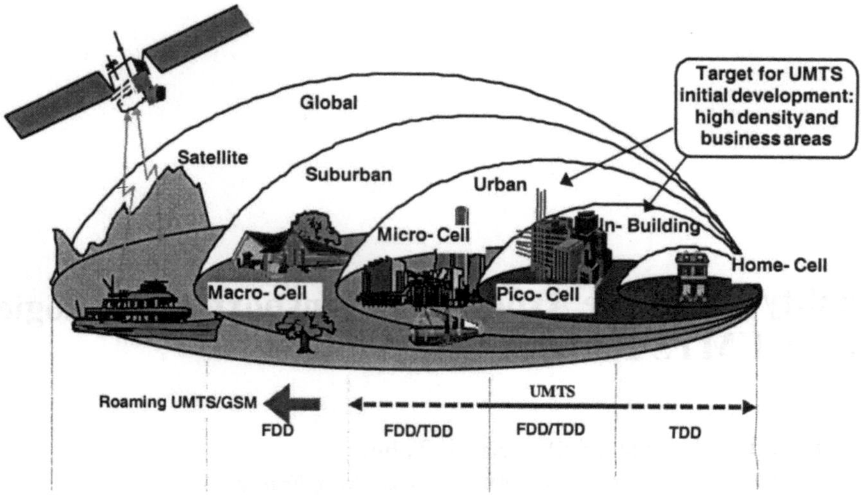

Figure 1: Scenarios for UMTS deployment

2. QOS IN UMTS AND THE RAN

The quality of service concept [5] developed in the 3rd Generation Partnership Project is based on the fundamental idea, that the user cares only about issues that are visible to him. As following this idea only the QoS perceived by end-users matters, it has to be provided end-to-end. Furthermore, the number of parameters to be defined or controlled by the user should be as small as possible, the derivation of QoS attributes from the application requirements should be simple and the QoS attributes should be able to support the asymmetric nature of some applications concerning the uplink and downlink direction.

The UMTS standardization is a continuos process and so the specifications are organised as a series of releases. The UMTS release '99 assumes the QoS related bearer definitions of GSM to be sufficient and aims to provide the services *speech*, *non-transparent data*, and *transparent data*.

The QoS Architecture proposed by 3GPP to support the QoS requirements of these services distinguishes different levels of QoS according to a layered architecture of *bearer services* in which services of one layer make use of the services of lower layers.

As user traffic passes from one terminal equipment (TE) to another it passes different bearer services of the network(s). The UMTS QoS architecture leaves aside the bearer services outside the scope of its' standardization and, therefore, concentrates on the UMTS bearer service and the bearer services this service is build upon.

Taking into account the restrictions and limitations of the air interface, which due to different error characteristics does not allow for definition of complex mechanisms like they have been proposed for fixed networks, the following four basic QoS classes are proposed:

a) The *conversational class* requires to preserve the time relation (variation) between information entities of a stream and puts stringent restrictions on the tolerable delay. The most well known example of this class is classical telephony, but with emerging use of Internet and multimedia a number of new applications, e.g. voice over IP or video conferencing tools, are expected to require this scheme.
b) The *streaming class* also demands for preservation of the timing relation between information entities of a stream but, as it tolerates higher end-to-end delays and thus gives more flexibility to align the timing relation at the receiver, the acceptable delay variation is much higher than that of the conversational class. The streaming class applies to unidirectional transport of information with the most well known applications being audio and video distribution.
c) The *interactive class* addresses the QoS requirements of applications where an end-user interacts with remote equipment in order to acquire some data. The fundamental QoS characteristics of this class are to preserve the payload contend and to meet an end-to-end delay that is suitable for the request-response pattern of its' applications, like web browsing or automated database queries.
d) The *background class* is supposed to be used for applications where the destination is not expecting data to arrive within a certain time, like this is the case with delivery of emails or content of push services, e.g. news. The only QoS requirement of this class is to preserve the payload content.

In order to be able to accurately describe the QoS requirements of traffic belonging to one of the above classes, the 3GPP is further working on the specification of suitable QoS parameters, which are still subject to ongoing discussions:

1. The *maximum bitrate* defines the maximum number of bits delivered by UMTS and to UMTS at a service access point within a period of time divided by the duration of the period. Traffic is conformant with the maximum bitrate as long as it follows a token bucket algorithm where the token rate equals the maximum bitrate and bucket size equals maximum SDU size.
2. The *guaranteed bitrate* specifies the guaranteed number of bits delivered by UMTS within a period of time, divided by the duration of the period. The conformance definition provides a parameter to specify a bucket size greater than the maximum SDU size in order to deal with bursty traffic.

However, for the 1999 release this parameter is fixed to 1, so that the bucket size equals the maximum SDU size.

3. The boolean parameter *delivery order* indicates whether the UMTS bearer service shall provide in-sequence delivery or not.
4. The specification of the *maximum SDU size* is needed for admission control and policing.
5. The *SDU format information* can be used to specify a list of possible SDU sizes, which allows for more efficient realization of the bearer service.
6. The *SDU error ratio* indicates the fraction of SDUs lost or detected as erroneous. The purpose of this parameter is to configure the protocols, algorithms and error detection schemes, primarily in UTRAN.
7. The *residual bit error ratio* indicates the undetected bit error ratio in delivered SDUs and is used to configure the radio interface protocols, algorithms, and error detection coding.
8. The parameter *delivery of erroneous SDUs* allows to specify if delivery of erroneous SDUs with or without an error indication, respectively, is wanted or not.
9. The *transfer delay* parameter indicates the maximum delay for 95th percentile of the distribution of delay for all delivered SDUs.
10. The *traffic handling priority* allows to specify the relative importance of the SDUs belonging to one bearer compared to the SDUs of other bearers. As priority handling is a fundamental alternative to absolute guarantees for providing QoS, this parameter can not be used together with parameters specifying absolute guarantees.
11. The *allocation / retention priority* specifies the relative importance of one bearer to other bearers and is used for call admission control in case of scarce resources. In the 1999 release this parameter is planned as a subscription parameter, which will not be negotiated with the user terminal.

These parameters are defined for both the UMTS and the radio access bearers and parameter value ranges are defined for each of them. Table 1 shows which parameter is relevant for each of the above QoS classes as well as the parameter ranges proposed so far for radio access bearers. Please note, that for radio access bearers there is one additional parameter *source statistics descriptor*, which permits to specify, that speech is the payload of a bearer, as this allows for calculation of a multiplexing gain for use in admission control of the radio access network.

Table 1. Radio access bearer attributes and value ranges for each QoS class

Traffic class	Conversational	Streaming	Interactive	Background
Maximum bitrate (kbit/s)	< 2048	< 2048	< 2048 - overhead	< 2048 - overhead
Delivery order	Yes / No	Yes / No	Yes / No	Yes / No
Maximum SDU size (octets)	<= 1502	<= 1502	<= 1502	<= 1502
SDU format information	to be done	to be done		
SDU error ratio	Yes / No / -	Yes / No / -	Yes / No / -	Yes / No / -
Residual bit error ratio	$5*10^{-2}$, 10^{-2}, $5*10^{-3}$, 10^{-3}, 10^{-4}, 10^{-6}	$5*10^{-2}$, 10^{-2}, $5*10^{-3}$, 10^{-3}, 10^{-4}, 10^{-5}, 10^{-6}	$4*10^{-3}$, 10^{-5}, $6*10^{-8}$	$4*10^{-3}$, 10^{-5}, $6*10^{-8}$
Delivery of erroneous SDUs	10^{-2}, $7*10^{-3}$, 10^{-3}, 10^{-4}, 10^{-5}	10^{-1}, 10^{-2}, $7*10^{-3}$, 10^{-3}, 10^{-4}, 10^{-5}	10^{-3}, 10^{-4}, 10^{-6}	10^{-3}, 10^{-4}, 10^{-6}
Transfer delay (ms)	80 (max.)	250 (max.)		
Guaranteed bitrate (kbit/s)	< 2048	< 2048		
Traffic handling priority			1,2,3	
Allocation / Retention priority	1,2,3	1,2,3	1,2,3	1,2,3

3. AAL-2 BASED TRANSPORT IN THE UTRAN

During the last couple of years the ITU-T has been working on a set of recommendations [13, 14, 15, 16, 17] defining an ATM adaptation layer that may be used for applications demanding a finer grained segmentation than that provided by existing adaptation layers, whose SDUs are at least 47 octets in size. The new adaptation layer AAL-2 allows for multiplexing of multiple communication streams into one or more ATM connections by multiplexing so called minicells, also referred to as AAL-2 *CPS-Packets*, containing a payload of 1 to 65535 octets into the cells of ATM connections. The 3GPP technical specification group for the radio access network (TSG RAN) has decided to use AAL-2 for transport between the radio network controllers (RNC) and Node Bs [1, 2, 3, 4] in order to allow for resource efficient transport of low-bitrate traffic in the radio access network.

AAL-2 is structured into multiple sublayers: the *common part sublayer (CPS)* and the *service specific convergence sublayer (SSCS)*, which itself is structured into the *service specific segmentation and reassembly (SSSAR)*, the *service specific transmission error detection (SSTED)*, and the *service specific assured data transfer (SSADT)* sublayer. Only the CPS needs to be present in an AAL-2 entity, the use of the other sublayers depends on the applications' needs.

The AAL-2 minicell header consists of three octets and starts with an eight bit *channel identifier (CID)*, allowing to multiplex up to 256 different AAL-2 connection into one ATM *virtual circuit (VC)*. The following *length indicator (LI)* is one less than the number of octets in the CPS-Packet payload. Its' maximum value may be fixed to 44 or 63, depending on the implementation (restricting the maximum to 44 may allow to re-use existing designs when building an AAL-2 switch). The field *user-to-user indication (UUI)* allows to signal 5 bits in the header of a minicell to the peer AAL-2 entity. It is followed by the field *header error correction (HEC)* which provides a 5 bit cyclic redundancy check code over the minicell header.

In order to insert minicells into the cells of an ATM-connection, AAL-2 entities assemble them into CPS-PDUs, each one filling exactly the payload of one ATM cell. The first field of a CPS-PDU (which is located directly after the ATM cell header) is the *offset field (OSF)* which carries the offset in octets to the first minicell header or the beginning of the pad field which is carried in this CPS-PDU where a value of zero indicates the position right after the *parity bit (P)*. In case, that there is neither a minicell header nor a pad field in this CPS-PDU it carries the value 47. The offset is used for minicell delineation. Between the offset field and the parity bit, a one bit *serial number (SN)* is located, which allows for detection of any uneven loss of CPS-PDUs. The payload of the CPS-PDU carries the minicells where minicells can arbitrarily cross the boundaries of CPS-PDUs within their length limits.

The ITU-T recommendation I.363.2 which standardizes AAL-2 doesn't deal in detail with questions related to QoS issues. However, it mentions to distinguish between different QoS-requirements of the AAL-2 user by the use of multiple service access points (SAPs, see also figure 2).

This situation gives rise to multiple questions concerning the realization of QoS in the context of the AAL-2 protocol:

- Which will be the QoS parameters or QoS classes offered to the service user of AAL-2 (i.e. definition of QoS at the AAL2 layer)?

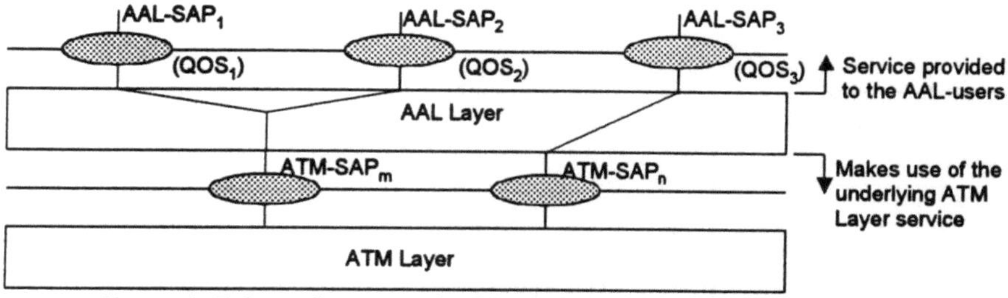

Figure 2: Relation between AAL-SAP and ATM-SAP (ITU-T Rec. I.363.2)

- What type of "guarantee" will be offered (statistic, as in the IP guaranteed service, but which is only provided as an indication associated with the connection, or deterministic, as in ATM on a per connection basis)
- Is there a need for a QoS negotiation or even re-negotiation at the AAL-2 layer, and if so, how might it be realized?
- How will this QoS specification be translated to the QoS offered by the ATM layer?
- What mechanisms can be used to ensure the QoS-guarantees given to the service user?

Basically, one has to distinguish between traffic parameters, that are descriptions of particular traffic aspects characterizing the traffic offered at a standardized interface, like peak cell rate, average cell rate, cell delay variation tolerances, burstiness, or peak duration, and quality of service (QoS) parameters, which characterize the transfer of traffic between two interfaces.

Concerning AAL-2 traffic parameters it seems reasonable to define them analogue to those defined for the ATM layer (cf. [19]): peak cell rate (PCR), cell delay variation tolerance (CDVT), sustainable cell rate (SCR), maximum burst size (MBS), etc. will have to be translated to peak minicell rate (PMR), minicell delay variation tolerance (MDVT), sustainable minicell rate (SMR), etc. Additionally, the AAL-2 traffic parameters should allow to specify the traffic streams properties concerning the size of exchanged minicells. Therefore, it seems obvious to add the mean and the maximum minicell size to the already existing traffic parameters.

In the same way as AAL-2 traffic parameters, the AAL-2 QoS parameters might best be defined analogous to the those of the ATM layer (see also [18] for detailed definitions), e.g. minicell error ratio (MER), minicell loss ration (MLR), minicell misinsertion rate (MMR), minicell transfer delay (MTD), and minicell delay variation (MDV).

It remains to be analyzed, how the guarantees given with an AAL-2 traffic contract can be assured by an AAL-2 entity. This is directly related to the means of traffic control deployed in the network, and particularly the realization of the scheduling, the buffer management as well as the bandwidth management in AAL-2.

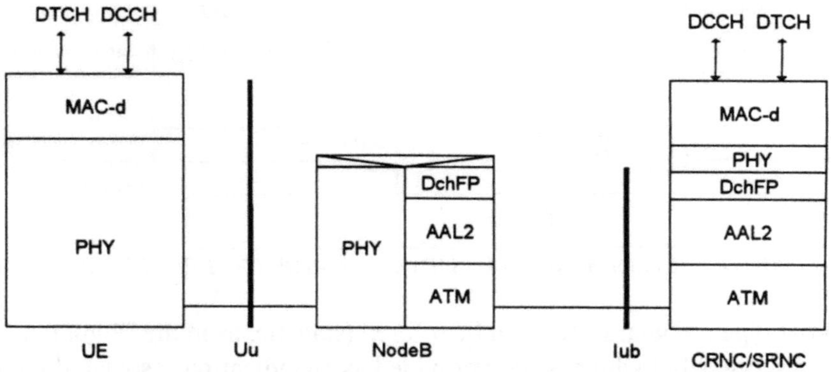

Figure 3: UTRAN protocol model for dedicated channels (3G TS 25.401)

However, when deployed in the UTRAN, this situation might be somehow simpler, as the protocol models of the UTRAN architecture reveal that the traffic to be transferred between two AAL-2 entities might be more homogenous, than the heterogeneous nature of the UMTS QoS classes seem to indicate at a first glance. Figure 3 depicts the protocol model for transport of dedicated channels between an RNC the Node Bs controlled by him.

As the RNC controls the radio resources of his Node Bs, the MAC layer processing, including the scheduling of heterogeneous communication streams, is handled by the RNC. This implies, that the purpose of a Node B and the corresponding AAL-2 connection to his RNC can be interpreted as "prolonging" the (upper part of the) physical layer between the mobile terminal and the RNC. In this respect, all traffic transported over AAL-2 in the UTRAN is in fact realtime traffic. This simplifies the constraints on AAL-2 in this scenario, as depending on the actual implementation it might not be necessary to provide sophisticated scheduling and buffer management mechanisms to cope with an arbitrary mix of realtime and non-realtime traffic.

The provision of quality of service in the transport network of the UTRAN has to serve two different purposes:

1. *Satisfy the user:* the requirements of the QoS classes defined for UMTS should met when transporting user data.
2. *Make good resource utilization:* this applies especially to the relatively rare and expensive radio resources and puts further constraints on the handling of traffic, as e.g. PDUs of the radio MAC need to reach the Node B on time, regardless of the QoS class of their content in order to avoid unused radio resources.

Summarizing, the answers to the questions, whether and how to distinguish between traffic with and without real time constraints on the level of AAL-2 will depend significantly on the design of the radio access network.

4. IP BASED TRANSPORT IN THE UTRAN

A quite recent development in the standardization of 3rd generation mobile communication systems is the investigation of an All-IP network architecture. This effort is conducted by the 3GPP2 consortium, which has been formed in reaction to the unwillingness of the ETSI to include other "non GSM" technologies in its' proposal for IMT-2000 [6].

The high level objectives of an All-IP network identified in a working report of 3GPP2 are [7]:

- to provide a unified (voice/data) wireless IP network that is interoperable with existing services, has gateways to legacy networks, and supports high capacity,
- to reuse existing radio network technology and to be independent of a specific air interface,
- to enable new services built on top of IP,
- to provide a global solution with an IP based infrastructure, as well as
- to maximize synergy and compatibility with existing standardization efforts.

Up to now, only vague ideas exist, how a 3rd generation mobile communication system based on an All-IP architecture should look like. Nevertheless, it is interesting to point out some of the QoS and performance issues of an All-IP based versus an ATM/AAL-2 based solution.

Various investigations [9, 11, 12, 21, 22, 23] are showing the performance advantages of AAL-2 for the transport of low bitrate voice data in comparison to prior ATM adaptation layers. Comparing AAL-2 to IP reveals firstly the differences in terms of header overhead. With AAL-2 every minicell comes with an overhead of 3 octets for up to 45/64 octets of payload plus additional 5 octets of overhead for every ATM cell, whereas IP can transport up to 65536 octets with only an overhead of 20 octets. So, IP will surely impose less overhead if the fraction of data traffic is high in comparison to voice traffic. However, overhead is just one performance aspect, as the ATM based approach might as well have difficulties to cope with very high data rates due to segmentation and reassembly processing.

Two architectures are currently considered. The first one is based on IP over MPLS with IETF differentiated services (DiffServ) as a framework of QoS. The second is based on native IP routing with DiffServ for the QoS.

5. ALTERNATIVE ARCHITECTURES FOR USRAN

UMTS involves a satellite segment which complements the terrestrial one to offer a global mobile communication service. The satellite component can play the same role of the access network which UTRAN presents in terrestrial UMTS segments and hence, will be referred to as UMTS Satellite Radio Access Network (USRAN). USRAN can be a backup option for terrestrial UTRAN in two cases. The first case arises in areas which are not covered by terrestrial networking infra-structures. The satellite component is then essential to relay the user equipment to the nearest UMTS or UMTS-compatible network. The second case occurs when the terrestrial UMTS segment is saturated, and no resources are available for arriving or handed over calls. The satellite segment can then act as an umbrella cell to provide extra resources to the terrestrial segment.

The mobility management in the satellite context is handled differently from the terrestrial case. Indeed, the cell size in the satellite case, several hundred kilometres, is much larger than in the terrestrial case (few kilometres). Thus the handover (HO) operations in the satellite case occur mostly because of the satellite mobility (the LEO/MEO case).

Two possible architectural scenarios can be imagined to integrate the satellite component in the UMTS access network. The first scenario sees the satellite acting as a Node B in the USRAN, while in the second, the space segment, consisting of one or more satellites, supports the dual functionality of a Node B and an RNC. In both cases, the uplink radio channel connects the user equipment to the UMTS core network through the space segment, which can be a geo-stationary satellite or a constellation of Low or Medium Earth Orbit satellites.

5.1 Satellite as Node B

The satellite segment in this architecture offers the functions of a Node B to the terrestrial UMTS segment. This scenario is most probable to encounter in areas with no terrestrial coverage and the satellite segment serves as a link between the user equipment and a remote RNC which is geographically out of reach of the user equipment. This implies that the satellite will act as a transceiver to support all the radio transmission functions, which are normally supported by the terrestrial Node B such as modulation, demodulation, rake receiver, coding etc. The radio transmission technology between the mobile and the satellite can be based on the ESA proposal [10] using Satellite Wideband CDMA SW-CDMA multiple access technique and operating in FDD mode with channel bandwidth of either 2.5 or 5 Mhz in each direction. The proposed solution is fairly close to the WCDMA used with UTRAN.

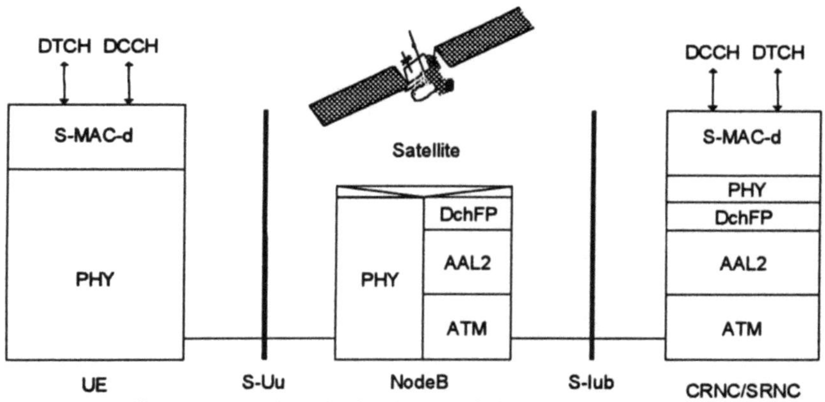

Figure 4: USRAN protocol model for dedicated channels with satellite as Node B

The particularity of the satellite link imposes additional precautions on the applied error detection and correction capabilities of the physical layer, between the terminal and the satellite in one hand and between the satellite and the RNC in the other hand, in order to keep the bit error rate comparable to the terrestrial segment and hence, preserve the quality of service parameters for the different services.

Figure 4 illustrates the protocol architecture when the satellite is playing the role of a Node B. The interface between user equipment and the satellite segment is referred to as S-Uu, while the interface between the satellite and the RNC is referred to as S-Iub. The AAL2 used within the S-Iub protocol stack may include sublayers SSTED for the error detection and SSADT (with SSCOP) for error correction in order to ensure reliable minicell exchange mode. The QoS provided by the access network to the user requirements can be defined at the MAC layer as a set of Transfer Capabilities MTCs. Three MTCs can be defined: CBR, Bursty Data and Best Effort [20].

5.2 Satellite as RNS

In this architectural scenario, the satellite segment can support the double role of an umbrella cell for terrestrial UMTS segments, as well as a relay between user equipment and remote UMTS core networks. The functions of both, the Node B and the RNC will be implemented in the satellite, which will then act as a spatial radio access network. The satellite segment will then handle the mobility, as well as admission and call control functions, besides all the physical aspects of transmission on both up- and down-links.

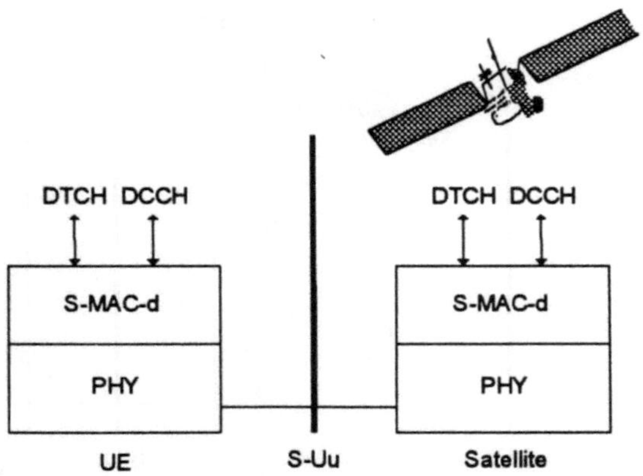

Figure 5: USRAN protocol model for dedicated channels with satellite as Node B

The protocol structure onboard the satellite is presented in figure 5. The satellite segment then ensures the functions of the physical, MAC, RLC, RRC and CC layers. The S-Uu interface relays the user equipment to the USRAN, while the S-Iu interface relays the space segment to the UMTS terrestrial core network. The satellite segment must be able to communicate directly with terrestrial RNCs, and hence the presence of the interface S-Iur. In addition, a satellite acting as USRAN must also be able to communicate directly with other satellites in the USRAN (in the presence of Inter-Satellite Links). Another interface intra-satellites must then be defined at this level. The MTCs are defined in the same manner than in the precedent architecture scenario.

5.3 Comparison

The first scenario is very similar with the UTRAN scheme. The complexity of the system in term of resource control is located within the terrestrial segment. The HO between terrestrial and spatial segments is simpler as this architecture corresponds to a hierarchical network. Because of the thorough control of satellite resources from the RNC, the satellite needs practically to be owned by the UMTS operator, which may reduce its economical feasibility.

In the second scenario the complexity is completely located on the segment space. This would allow for a global satellite access operator who offers his services to multiple UMTS operators. In this case, however, the HO will correspond to a more complex roaming operation.

6. CONCLUSION

The radio access network technology of the 3rd generation mobile networks are anticipated to evolve towards a packet oriented architecture, as it is expected that the fraction of data traffic will increase in relation with mobile computing power increase. The ATM technology with AAL-2 might be seen as a first step in the 3rd generation radio network evolution as it is build on the premise, that low bitrate voice transmission will be the dominant part of the traffic. IP technology including the ongoing work of the IETF on IP QoS is currently appearing as the next technological step to operate within fixed/mobile and terrestrial/satellite networks.

REFERENCES

[1] 3GPP Technical Specification Group Radio Acces Network (TSG RAN). *UTRAN Overall Description.* 3G TS 25.401.

[2] 3GPP TSG RAN. *UTRAN Iu Interface, Data Transport and Transport Signalling.* 3G TS 25.414.

[3] 3GPP TSG RAN. *UTRAN Iur and Iub Interfaces, Data Transport and Transport Signalling for DCH Data Stream.* 3G TS 25.426.

[4] 3GPP TSG RAN. *UTRAN Iub Interface, General Aspects and Principles.* 3G TS 25.430.

[5] 3GPP TSG Services and System Aspects. *Qos Concept and Architecture (Release 1999).* 3G TS 23.107.

[6] 3GPP2. *Background on the 3rd Generation Partnership Project 2.* http://www.3gpp2.org/text/background.cfm, 1999.

[7] 3GPP2 TSG-S. *All IP Adhoc Report to TSG-S – 01/00.* S00ALLIP-20000106-024, January 2000.

[8] A. Adas. *Traffic Models in Broadband Networks.* Communications Magazine, pp. 82-89, IEEE, July 1997.

[9] J. H. Baldwin, B. H. Bharucha, B. T. Doshi, S.. Dravida, S. Nanda. *AAL 2 – A New ATM Adaptation Layer for Small Packet Encapsulation and Multiplexing.* Bell Labs Technical Journal, Spring 1997.

[10] ESA. *Wideband CDMA Option for the Satellite Component of IMT-2000 "SW-CDMA".* ESA Proposal for a Candidate RTT. V. 1.0. June 1998.

[11] N. Gerlich, M. Ritter. *Carrying CDMA Traffic over ATM using AAL-2: A Performance Study.* Research Report No. 188, Department of Distributed Systems, University of Würzburg, November 1997.

[12] N. Gerlich, M. Menth. *The Performance of AAL-2 Carrying CDMA Voice Traffic.* Research Report No. 199, Department of Distributed Systems, University of Würzburg, April 1998.

[13] ITU-T. *B-ISDN ATM Adaptation Layer specification: Type 2 AAL.* Recommendation I.363.2, 1997.

[14] ITU-T. *Segmentation and Reassembly Service Specific Convergence Sublayer for the AAL type 2.* Recommendation I.366.1, 1998.

[15] ITU-T. *AAL Type 2 Service Specific Convergence Sublayer for Narrow-Band Services.* Draft Recommendation I.366.2, March 2000.

[16] ITU-T. *AAL Type 2 Service Specific Convergence Sublayer for Mobile.* Draft Recommendation I.366.3, November 1999.

[17] ITU-T. *AAL Type 2 Signalling Protocol (Capability Set 1).* Recommendation Q.2630.1, 1999.

[18] ITU-T. *B-ISDN ATM Layer Cell Transfer Performance.* Recommendation I.356, October 1996.

[19] ITU-T. *Traffic Control and Congestion Control in B-ISDN.* Recommendation I.371, August 1996.

[20] H. Koraitim, S. Tohmé. *Resource Allocation and Connection Admission Control in Satellite Networks.* IEEE Journal on Selected Areas in Communications, February 1999.

[21] C. Liu, S. Munir, R. Jain, S. Dixit. *Packing Density of Voice Trunking Using AAL 2.* Globecom '99, Rio de Janeiro, Brazil, December 1999.

[22] K. Sriram, T. G. Lyons, Y. T. Wang. *Anomalies Due to Delay and Loss in AAL-2 Packet Voice Systems: Performance Models and Methods of Mitigation.* IEEE Journal on Selected Areas in Communications, January 1999.

[23] K. Sriram, Y. T. Wang. *Voice over ATM using AAL-2 and Bit Dropping: Performance and Call Admission Control.* IEEE Journal on Selected Areas in Communications, January 1999.

Resource Allocation in a Cellular CDMA Environment

R. Bolla, F. Davoli, M. Perrando
Department of Communications, Computer and Systems Science (DIST), University of Genoa, Via Opera Pia 13, I-16145 Genoa, Italy
{lelus, franco, perr}@dist.unige.it

Key words: Call admission control, CDMA, resource allocation, fixed point, cellular systems.

Abstract A cellular environment is considered, based on CDMA. The main goal is to optimize the overall system capacity, in order to equalize and minimize the blocking probability of new calls entering into the system; at the same time, a constraint on the outage probability and, possibly, another one on handoff blocking are enforced. The control is based on the multiple access interference threshold, and the optimization algorithm tries to find the best choice of the admittance thresholds, based on the "a priori" knowledge of the traffic characteristics, and of a mobility model of the users.

1. INTRODUCTION

The growing user demand and the diversification of services in mobile wireless networks are stimulating the deployment of systems with smaller cell size and higher potential flexibility in the allocation of the resources, both across different services and among groups of cells. Such dynamic resource allocation aims at maintaining Quality of Service (QoS) requirements, even in the presence of traffic flows with different statistical nature and performance characteristics and of variable user mobility patterns. Notwithstanding the multiple access method adopted, some basic factors that play a key role in the QoS provisioning to multimedia traffic are the adaptation to user mobility, the provision of differentiated (per-traffic-class) resource allocation and the overall system's organization into coordinated management domains. In this respect, control actions can be applied at different operational layers in the network structure.

In particular, bandwidth allocation can be considered at the channel level, in conjunction with the multiple access protocol, in different multiple access (TDMA or CDMA) environments (see, e.g., [1], [2], [10], [5]), or at a "higher" hierarchical layer, affecting the network structure and in conjunction with Call Admission Control (CAC); see [12], [9], [3], [4], among others. A control architecture based on a hierarchical paradigm, which reflects a possible network organization in cells and cell clusters, has been considered by the authors in a pure TDMA context [3], [4]. In this paper we consider a similar control structure, but embedded in a CDMA scheme, which is the preferred choice in the emerging UMTS [11].

We combine two mechanisms that operate at different levels within the network: i) a threshold-based admission control within the cell, which sets different acceptance levels for new calls originating in the cell and handoff calls coming from adjacent ones; ii) a dynamic resource sharing scheme at the cluster level, for the distribution of the system's capacity (in terms of the maximum number of calls compatible with a given multiple access interference) among the cells in a cluster.

The paper is organized as follows. We describe our model of the CDMA channel in the next section. In Section 3, we introduce the resource allocation mechanism and a numerical approximation to compute the Multiple Access Interference (MAI) with a given set of thresholds. In Section 4, we report and discuss several numerical results deriving from the application of the proposed scheme. Section 5 contains the conclusions.

2. THE MODEL OF THE CELLULAR ENVIRONMENT

In the following we give a specification of the environment considered in the present work, and also some hypotheses that have been taken into account in order to simplify calculations. A very similar model has been used in [6].

A cellular environment has been considered, in which Base Stations (BS's) perform a perfect power control on the Mobile Stations (MS's) under their domain; furthermore, exogenous (environmental, Gaussian, etc.) noise has been ignored, with respect to the endogenous interference. Voice activity detection has not been adopted in the present version of the work: a mobile transmits continuously for the whole duration of its call.

The total area is divided into C hexagonal cells all of equal size, and the BS's antenna is omnidirectional and placed in the center of the cell.

A MS is controlled by the BS which has the best radio propagation factor with that MS.

There is a complete separation between the forward and the reverse link, so that there is no interference between BS's and MS's.

The radio propagation model assumed is a generally accepted one [7], where the total attenuation depends on two variables: a random shadowing factor, log-normal distributed, and a path loss factor, depending on the α power of the distance. In other words, the attenuation factor suffered by a MS at distance r from a BS is:

$$\theta(r) = 10^{\xi/10} r^{-\alpha} \tag{1}$$

where ξ is a Gaussian random variable with standard deviation σ. The values for σ and α generally adopted, and also considered in the present work are $\sigma = 8$, and $\alpha = 4$.

The new call (not originated by MS movements) arrival process is modeled, at a generic cell i, as an independent Poisson process with mean arrival rate λ_i^{nc}, whereas the mean call lifetime is exponentially distributed with mean duration of $1/\mu$.

The model of MS mobility is analogous to [4]: a mobility coefficient γ_i, depending on the cell site, gives the fraction of calls that leave cell i because they handed-off to another site, as explained more in detail in the following. At the boundary of the system a hand-off flow of entering calls is also considered.

The admission scheme is based on the measure of the Multiple Access Interference Ratio (MAI) estimated by every BS, supposed to be measured exactly. The admission control policy may operate differently, depending on the fact that the call requesting admission is generated by a hand-off situation, or it is a new call entering the system. Calls are admitted into the system if the instantaneous MAI is under a determined level. This MAI threshold may be different for the hand-off case and the new call case.

Given the parameters of the spread spectrum modulation, such as the processing gain, we can estimate the maximum value for the MAI above which the quality of the transmission falls under a minimum requisite. Let MAI_0 be this value. The probability that the cell MAI overcomes MAI_0 will be called outage probability.

3. THE ALLOCATION PROCEDURE

The main goal of this work is to maintain the outage probability in every single cell of the system under a given constraint $\hat{P}^{otg}$. In addition to this, another constraint $\hat{P}^{b,ho}$ may be imposed over the blocking probability of the calls entering for hand-off. This last constraint may be applied only in the case of different thresholds for hand-off calls and new calls.

Let us see in detail how the MAI is calculated over the system. If k is the number of active calls in cell i, the total power $I(i)$ received by BS i is given by:

$$
\begin{aligned}
I(i) &= \sum_{u \in \{\text{all MS's}\}} I_u(i) = \\
&= Sk + S \sum_{j \neq i} \sum_{u \in \{\text{MS's not in cell } i\}} \left(\frac{r_j^u}{r_i^u}\right)^{\gamma} 10^{(\xi_i^u - \xi_j^u)/10} = \\
&= I_{int}(k) + I_{ext}(i)
\end{aligned}
\tag{2}
$$

where $I_u(i)$ is the power of the signal received by BS i from a MS u transmitting to its own BS j; r_j^u and ξ_j^u are the distance and the shadowing, respectively, of MS u with respect to its own BS j; r_i^u and ξ_i^u are the distance and the shadowing, respectively, of MS u with respect to BS i; and S is the power received by a BS from a perfectly power-controlled MS. Furthermore, $I_{int}(k)$ is defined as the contribution originated by k MS's directly controlled by a BS, while $I_{ext}(i)$ is defined as the interference caused on BS i by all the MS's not controlled by that BS.

The MAI measured at BS i, having k active calls, thus results:

$$
MAI(i,k) = \frac{I(i) - S}{S} = k - 1 + MAI_{ext}(i) \tag{3}
$$

where we have defined

$$
MAI_{ext}(i) = \frac{I_{ext}(i)}{S} \tag{4}
$$

Even assuming a spatially uniform distribution of the users over the ground, each of the variables $MAI_{ext}(i)$, clearly depends on the global status of all the cells in the system; in other words, it depends on the number of calls being active in every cell, or, more precisely, on their joint distribution. In order to approximate the MAI of a given cell, with a quantity dependent only from the number of calls active in such cell, we remove the dependence on the status of other cells in the system, by considering each variable $MAI_{ext}(i)$ as a stochastic variable, having its own distribution. This distribution is then evaluated, assuming to know the steady state probability of the number of active calls in every cell.

In order to calculate the distribution of $MAI_{ext}(i)$, we consider the contribution to the MAI suffered by a generic BS i, given by a single

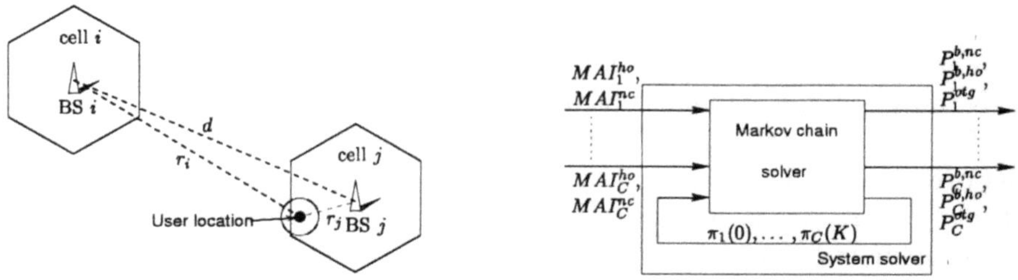

Figure 1 (a) MAI of a single user; (b) Fixed point solution

active call, which is running in a cell under the control of BS j. We call $MAI_{single}(d)$ this contribution, because it depends only on the relative distance d between the two BS's, and its value is given by [7]:

$$MAI_{single}(d) = \frac{1}{A} \int_{\text{cell } j} \mathcal{M}(r_i, r_j) dA \tag{5}$$

where A is the area of a cell, r_i and r_j are the distance of the area dA from BS i and BS j, respectively (see Figure 1 (a)), and

$$\mathcal{M}(r_i, r_j) = \left(\frac{r_j}{r_i}\right)^{\alpha} \left\{10^{(\xi_i - \xi_j)/10}\right\} \Phi\left(\xi_i - \xi_j, \frac{r_j}{r_i}\right) \tag{6}$$

where

$$\Phi\left(\xi_i - \xi_j, \frac{r_j}{r_i}\right) = \begin{cases} 1, & \text{if } (r_j/r_i)^{\alpha} 10^{(\xi_i - \xi_j)/10} \leq 1 \\ 0, & \text{otherwise} \end{cases} \tag{7}$$

Now, as in [7], we consider the $MAI_s(d)$ as a **Gaussian variable** having its own mean and variance. Then, following an analogous procedure, we numerically calculate (from (5)) both the mean value and the variance.

As already stated, we suppose to know the steady state probability of the number of active calls in the system. $\pi_j(k)$ being the probability of having k active calls in the generic cell j, we can write $MAI_{ext}(i)$ as

$$\begin{aligned} MAI_{ext}(i) &= \sum_{j \neq i} \sum_{k=0}^{K} MAI_{single}(d_{i,j}) k \pi_j(k) = \\ &= \sum_{j \neq i} MAI_{single}(d_{i,j}) \bar{n}_j \end{aligned} \tag{8}$$

where $\bar{n}_j$ is the average number of active calls in cell j, $d_{i,j}$ is the distance between the two BS's, and K is a "large enough" value for the maximum number of active calls that can be present in every cell, so that the probability of having more then K active calls is zero. A good choice for K may be the capacity of a single isolated cell: due to non null interference of other cells, the CAC cannot bring the number of active calls over this value.

Since $MAI_{ext}(i)$ is the sum of Gaussian distributed variables, it has a Gaussian distribution having as mean value the sum of mean values of the factors, and, as the variance, the sum of the variances of the factors. Let the outage probability suffered by cell i be

$$w_i^{otg}(k) = \mathbf{Pr}\{MAI(i,k) > MAI_0\} \tag{9}$$

The control variables that decide the admission of calls are defined as:

$$q_i^{nc}(k) = \begin{cases} 1 & \text{if } MAI(i,k+1) < MAI_i^{nc} \\ 0 & \text{otherwise} \end{cases} \tag{10}$$

representing the acceptance for a new call if $q_i^{nc} = 1$ at cell i and

$$q_i^{ho}(k) = \begin{cases} 1 & \text{if } MAI(i,k+1) < MAI_i^{ho} \\ 0 & \text{otherwise} \end{cases} \tag{11}$$

representing the acceptance of a call entering cell i for hand-off if $q_i^{ho} = 1$.

Defining

$$w_i^{ho}(k) = \mathbf{Pr}\{q_i^{ho}(k) = 1\} \tag{12a}$$

$$w_i^{nc}(k) = \mathbf{Pr}\{q_i^{nc}(k) = 1\} \tag{12b}$$

then, the total load entering the generic cell i is

$$\lambda_i(k) = \lambda_i^{ho} w_i^{ho}(k) + \lambda_i^{nc} w_i^{nc}(k) \tag{13}$$

where the quantities λ_i^{nc} represent the arrival rates of the calls originated by a new generation (not by hand-off), and λ_i^{ho} represent the rates of calls that enter the cell for hand-off. These quantities are reduced, by the effect of the CAC policy, of a factor $w_i^{nc}(k)$ and $w_i^{ho}(k)$, respectively.

While the quantities λ_i^{nc}, $i = 1, \ldots, C$ are given data of the problem, each λ_i^{ho} depends on the state of the whole cellular environment, as well as on the mobility model.

Under the previous assumption we can calculate the acceptance factors in (10) and (11), and, consequently, the fraction of the accepted calls in (12). But now, in order to evaluate $\lambda_i(k)$ in (13), we need also the quantities λ_i^{ho}. The latter may be found from the mobility model of users, which will be now explained in detail.

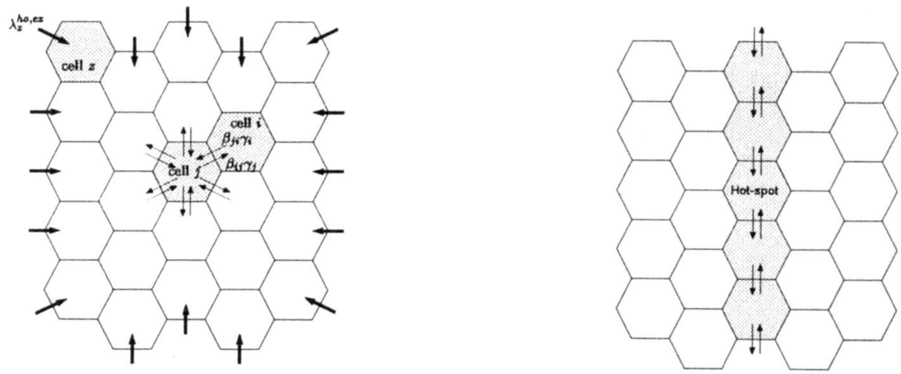

Figure 2 (a) Mobility model; (b) Optimization scenario

As already stated, every cell i has a mobility factor γ_i, that indicates the rate of calls that die for hand-off. Being μ the mean call death rate for "natural termination", we can state that, given k active calls within a cell, the call death rate is, for that cell, $k\mu$. The role of γ_i is to state the call death rate for hand-off; more precisely, this rate is, for cell i, having k active calls,

$$\mu_i^{ho}(k) = \gamma_i \mu k \tag{14}$$

By this, the total death rate for cell i having k active calls is

$$\mu_i(k) = (1 + \gamma_i)\mu k \tag{15}$$

Furthermore, we need to know what is the part of the total exiting calls that goes to each of the neighboring cells. The part of traffic that exits from cell j and goes to the adjacent cell i is given by the coefficient β_{ij}, for which the following equations must be satisfied

$$\beta_{ij} = 0 \quad \text{if } i \notin \mathcal{A}(j) \tag{16a}$$

$$\sum_{i \in \mathcal{A}(j)} \beta_{ij} = 1 \tag{16b}$$

where $\mathcal{A}(j)$ is the set of cells adjacent to cell j.

An exogenous hand-off rate deriving from users that enter the cell cluster from outside is also considered for every border cell. Thus, for every cell i, an additional rate of $\lambda_i^{ho,ex}$ is taken into account, with the caution to set $\lambda_i^{ho,ex} = 0$ if i is not a border cell.

A scheme for mobility is illustrated in Figure 2 (a), where the behavior of one cell is detailed.

The total rate of calls entering cell i for hand-off is thus:

$$\begin{aligned}\lambda_i^{ho} &= \sum_{j\in\mathcal{A}(i)} \beta_{ij} \sum_{k=0}^{K} \mu_j^{ho}(k) k \pi_j(k) + \lambda_i^{ho,ex} = \\ &= \sum_{j\in\mathcal{A}(i)} \beta_{ij}\gamma_j \mu \bar{n}_j + \lambda_i^{ho,ex}\end{aligned} \tag{17}$$

where $\bar{n}_j = \sum_{k=0}^{K} k\pi_j(k)$ is the mean value of the number of active calls in cell j.

Now we have all the elements ($\lambda_i(k)$ and $\mu_i(k)$) to solve the Markov chain for every cell i and to find out the steady state probability $\pi_i(k)$ of having k active calls in cell i.

In other words, we are trying to solve a vectorial fixed point equation in the form:

$$x = f(x) \tag{18}$$

where $x = \mathbf{col}[\pi_1(0), \ldots, \pi_C(K)]$.

Due to the very high complexity of function $f(\cdot)$, it is impossible to solve explicitly such an equation. As in [4], the assumption of existence of a single fixed point is made. This assumption is also supported from the work of [8].

Starting from a random value for the $\pi_i(k)$, we may calculate all the values needed in order to recompute these probabilities. Repeating these steps until some norm of the $\pi_i(k)$ converges, gives rise to the solution of the equation. A graphical scheme illustrating these procedure can be found in Figure 1 (b). Now we can calculate the blocking probabilities of new calls and hand-off calls, and the outage probability of every cell:

$$P_i^{b,nc} = \sum_k (1 - w_i^{nc}(k))\pi_i(k) \tag{19a}$$

$$P_i^{b,ho} = \sum_k (1 - w_i^{ho}(k))\pi_i(k) \tag{19b}$$

$$P_i^{otg} = \sum_k w_i^{otg}(k)\pi_i(k) \tag{19c}$$

The variables that remain fixed during this process are the admission thresholds MAI_i^{nc} and MAI_i^{ho}. These are the control variables that we can use to optimize the capacity among different cells. The optimization goal is to minimize a global cost represented by the maximum blocking probability of the new calls

$$P^{b,nc} = \max_i \{P_i^{b,nc}\} \tag{20}$$

In other words, we want to solve the following problem

$$\min_{MAI_1^{nc}, MAI_1^{ho}, \ldots, MAI_C^{nc}, MAI_C^{ho}} \{P^{b,nc}\} \tag{21}$$

In this scenario we set a constraint over the quality, imposing a minimum level on the quality of the call at "transmission level", by requiring that the outage probability must remain under a certain threshold:

$$P^{otg} = \max_i \{P_i^{otg}\} \leq \hat{P}^{otg} \tag{22}$$

An additional analogous constraint may be imposed over the blocking probabilities of hand-off calls, in order to ensure a "call-level" quality for calls that are running. In such a case, we set the following constraint

$$P^{b,ho} = \max_i \{P_i^{b,ho}\} \leq \hat{P}^{b,ho} \tag{23}$$

Otherwise, we may treat hand-off and new call connections equally. In this case, there is no difference between new or hand-off calls; thus, both are considered as new calls, and only the value of $P^{b,nc}$ is taken into account, paying attention to the fact that, in this case, it represents the blocking probability of both new and hand-off calls.

4. OPTIMIZATION RESULTS

We call **case A** the case for which no constraint on the hand-off blocking probability is set, and an identical CAC policy is adopted for both kinds of connections, and **case B** the one for which we set a constraint on the maximum rate of blocked hand-off calls. In the latter the maximum blocking probability of 10^{-3} is imposed.

The optimization consists of solving a constrained mathematical programming problem with non-differentiable cost function, over a domain in a $2C$-dimensional space, in case of different thresholds for hand-off and new calls, and over a C-dimensional space in the other case.

The procedure adopted consists of a kind of genetic algorithm, that generates a new population, taking into account a heuristic that tries to optimize the cost on every single cell.

The scenario we take into account consists of a 5×5 cell set, within which the possible hot-spot cell is the central one. Furthermore, the presence of a traffic lane is considered in the sense of the arrows. The latter is modeled with both an increment of users mobility and new call generation rate. All this can be depicted as in Figure 2 (b).

There are many parameters that can change the final result of the optimization, such as: mean load of the system, presence or absence of a hot-spot cell, mobility factor of the users, presence of a high hand-off rate in one predominant direction, etc.

Several cases have been taken into account, for which, unless otherwise indicated, the mobility coefficient is $\gamma_i = 0.3\ \forall i$, the fraction of users that exit towards every adjacent cell is $\beta_{ij} = 1/6$, the "transmission level" constraint is $\hat{P}^{otg} = 10^{-3}$, the single isolated cell capacity is $MAI_0 = 90$, and the exogenous rate for a border cell z is $\lambda_z^{ho,ex} = 0.25 l_z$, where l_z are the number of border sides of cell z. In all results presented below, the constraint on the outage probability is respected.

The first configuration optimizes the allocation in a homogeneously loaded system with increasing load. Numerical results of the computation are plotted in both A and B cases, and are represented in Figure 3 (a), where blocking probabilities are plotted against the cell traffic load. The numerical scale of the hand-off blocking probability in case B is mapped on the right of the plot for a scaling problem. It is clear how, in case B, the constraint over hand-off call blocking produces a slight increase in new call blocking probability.

Furthermore, the presence of a single hot-spot cell, having an increased new call generation rate, within a lightly loaded system, has been considered. The remaining cells present a homogeneous load of 20 Erlangs. Plots for an increasing traffic load of the hot-spot cell are presented, considering cases A and B. In Figure 3 (b) blocking probabilities are displayed as a function of the hot-spot cell load.

Another case of a single hot-spot cell has then been taken into account. Here the other cells in the system are homogeneously loaded at 50 Erlangs. Again, plots for an increasing traffic load of the hot-spot cell are presented, considering case A and B. In Figure 3 (c) blocking probabilities are displayed as a function of the hot-spot cell load.

Finally, the plot of blocking probabilities referring to the traffic lane is plotted in Fig 3 (d), again for both A and B cases, against an increasing load of the cells belonging to the traffic lane. The mobility coefficient relative to such cells has been increased to $\gamma_i = 0.4$, and also the coefficients β_{ij} of the sides marked by arrows in Fig 2 (b) are slightly increased.

5. CONCLUSIONS

A CDMA wireless cellular network has been considered, where the maximum capacity is fixed by a given level of multiple access interference. This global resource is dynamically assigned to different cells, according to the changing user mobility patterns and load variations. The assignment strategy is based on a two-level control, where a set of

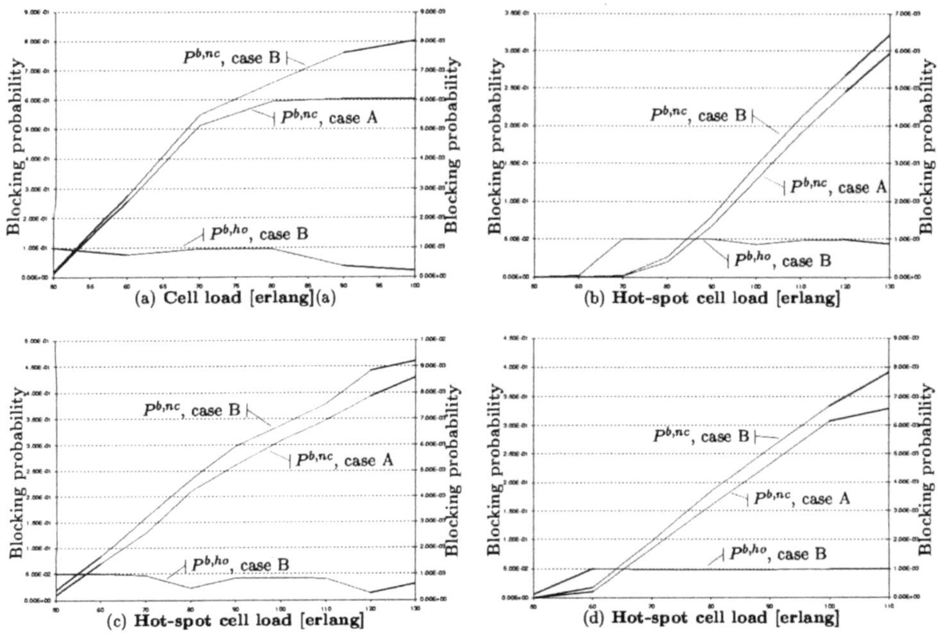

Figure 3 Results(read $P^{b,ho}$ values on the right scale) for: (a) homogeneous load (b) hot-spot "unloaded" (c) hot-spot "loaded" (d) traffic lane

thresholds implement the required resource distribution, whereas a controller within each cell decides on the admission of new calls. Numerical results show the effectiveness of the mechanism in keeping the handoff rejection level below a given upper bound, while always respecting the constraints on the outage probability.

REFERENCES

[1] Anastasi G., D. Grillo, L. Lenzini "An access protocol for speech/data/video integration in TDMA-based advanced mobile systems" *IEEE J. Select. Areas Commun.*, vol. 15, no. 8, pp. 1498-1510, Oct. 1997.

[2] Bolla R., F. Davoli, C. Nobile "The RRA-ISA multiple access protocol with and without simple priority schemes for real-time and data traffic in wireless cellular systems" *Mobile Networks and Applications*, vol. 2, pp. 45-53, 1997.

[3] Bolla R., D. Budicin, F. Davoli, R. Rossi "A coordinated structure for call admission control and dynamic channel allocation in wireless cellular networks" *Proc. IEEE Internat. Symp. on Wireless Commun. (ISWC'99)*, Victoria, BC, Canada, June 1999 (Extended Abstract).

[4] Bolla R., F. Davoli "An iterative optimization scheme for resource adaptation to user mobility patterns in cellular wireless systems" to be presented at *IEEE Wireless Communications and Networking Conference 2000 (WCNC'2000)*, Chicago, IL, Sept. 2000.

[5] Liu T.K., J.A. Silvester "Joint admission/congestion control for wireless CDMA systems supporting integrated services" *IEEE J. Select. Areas Commun.*, vol. 16, no. 6, pp. 845-857, Aug. 1998.

[6] Liu Z., M. El Zarki "SIR-Based Call Admission Control for DS-CDMA Cellular Systems" *IEEE J. Select. Areas Commun.*, vol. 12, no. 4, pp.638-644, May 1994.

[7] Gilhousen K.S., I.M. Jacobs, R. Padovani,A.J. Viterbi, L.A. Weaver, Jr., C.E. Wheatley III "On the Capacity of a Cellular CDMA System" *IEEE Trans. Vehic. Technol.*, vol. 40, no. 2, pp.303-311, May 1991.

[8] Mainkar V., K.S. Trivedi "Sufficient conditions for existence of a fixed point stochastic reward net-based iterative models" *IEEE Trans. Softw. Engin.*, vol. 22, no. 9, pp. 640-653, Sept. 1996.

[9] Naghshineh M., A. S. Acampora "QoS provisionig in micro-cellular networks supporting multiple classes of traffic" *Wireless Networks*, vol. 2, pp. 195-203, 1996.

[10] Sampath A., J.M. Holtzman "Access control of data in integrated voice/data CDMA systems: benefits and tradeoffs" *IEEE J. Select. Areas Commun.*, vol. 15, no. 8, pp. 1511-1526, Oct. 1997.

[11] Samukic A. "UMTS Universal Mobile Telecommunications System: Development of standards for the third generation" *IEEE Trans. Vehic. Technol.*, vol. 47, no. 4, pp. 1099-1104, Nov. 1998.

[12] Yu O.T.V., V.C.M. Leung "Adaptive resource allocation for prioritized call admission over an ATM-based wireless PCM" *IEEE J. Select. Areas Commun.*, vol. 15, no. 7, pp. 1208-1225, Sept. 1997.

An Improved Speech and Channel Coding for GSM System

Dariusz Godyń, Dominik Rutkowski
Motorola Polska Software Center, ul. Jodłowa 3, 30-252 Kraków, Poland
e-mail: Dariusz.Godyn@motorola.com
Department of Radio Communication Systems, Technical University of Gdańsk
e-mail: nick@pg.gda.pl

Key words: GSM system, channel error protection

Abstract: In the paper an improved speech and channel coding/decoding for the GSM system has been described. It enables to decrease the source rate and to increase the channel error protection of the speech and thus to achieve better performance of the reconstructed speech than the original speech and channel coding/decoding in the presence of fading and noise.

1. INTRODUCTION

The major task of the source coding/decoding of speech is to reduce the redundancy of speech and the corresponding bit rate in order to confine the required channel bandwidth that is a valuable resource in any radio communication system. On the other hand, the channel coding/decoding that introduces some redundancy is used to increase the error protection of speech data transmitted over the channel with fading, interference and noise to improve the performance of the reconstructed speech at the receiver. Despite of that protection one may observe the loss of speech frames in GSM system at the poor channel conditions and, of course, a drastic degradation of performance. However, if one looks more closely on the operation of source coder in GSM system one can notice the possibility of additional reduction of bit rate at its output that combined with more robust error protection on the channel can improve the overall performance of speech reconstruction at the output of the receiver if compared with the standard solution.

2. FULL-RATE CODING WITH REDUCED BIT RATE

The source coding of speech in GSM system is based on the digital model of a natural mechanism of speech signal generation. The essence of source coding consists in periodic evaluation of specific parameters characterizing the digital model provided that the segment of source signal is taken in the sufficiently short time interval in which it can be considered as quasistationary. The set of these parameters called a vector is used in the source decoder at the receiver to the synthesis of speech signals. The digital model of speech signal generation shown in Fig. 1 consists of the excitation generator, long- and short-term prediction filters as well as a unit that is approximating each source signal segment by the synthesized speech signal segment to obtain the weighted error minimization. All the parameters characterizing the excitation generator and the prediction filters are quantized usually with the nonlinear quantizers and periodically transmitted over the channel.

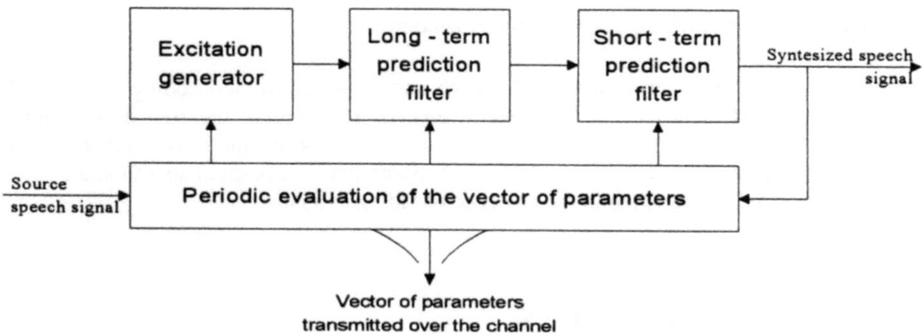

Figure 1. General model of the evaluation of parameters characterizing source speech signal *used in GSM system*

The excitation generator is modelling a flow of air stream through the vocal tract of a human speech organ. The linear long-term prediction filter is modelling the pitch period corresponding to the vibrations of vocal cord. The linear, short-term prediction filter is modelling the short-term spectral envelope of speech.

The number of bits delivered periodically at the output of speech encoder depends on the number of parameters describing the model of the source signal and the range of their values that in turn depend on the model complexity and the required performance of synthesized speech.

The standard speech encoder employed in GSM system is based on the RPE- LTP- LPC algorithm [1]. It produces the bit rate of 13 kb/s which when compared with the input bit rate of 104 kb/s gives eightfold compression at the acceptable quality of the reconstructed speech in terms of MOS value (MOS ~3.6÷4.0) [2], [3]. This result is obtained at the moderate complexity in terms of operations per second and short delay due to processing. Speaking more precisely, the GSM speech encoder delivers for each 20ms segment of speech samples, 8 parameters of short- term prediction filter that are represented by 36 bits and 2 parameters (gain and lag) of long- term prediction filter that are taken every 5ms and coded by 2 and 7 bits, respectively which yields another 36 bits. The parameters of excitation generator are calculated 4 times within one speech segment. These are 13 excitation pulses and the initial time phase of their appearance coded by 2 bits. The amplitudes of excitation pulses are first normalized with respect to the largest of them and then the largest amplitude is coded by 6 bits and the other ones by 3 bits. Summing up, the parameters of excitation pulses are represented by 188 bits every 20 ms and the speech encoder produces 260 bits of speech frame for each segment of source speech signal. The position of each bit in that frame depends on its weight i.e. its influence on the intelligibility of reconstructed speech. According to that weight 260 bits of speech frame are subdivided into 3 classes as it is shown in Fig.2 with the most important class 1a of 50 bits, less important class 1b of 132 bits and the least important class 2 of 78 bits. There is different channel coding scheme used for bits class 1a and 1b.

The speech decoder at the receiver is synthesizing the speech signal on the basis of the received speech frame corrupted by noise, fading and interference on the channel.

As it turns out the number of 260 bits contained in the speech frame can be reduced by using a procedure for finding out optimized quantizers of parameters for the source signal model [4]. The characteristics of quantizers have been obtained with the use of Max – Lloyd algorithm and the kernel estimator of the probability density functions for different parameters.

The investigations carried out with respect to the subjective audio tests of reconstructed speech and the analysis of the average parameter-to-quantization noise ratio (PQR) for each standard quantizer enabled to find some new optimum quantizers. They concern 8 parameters of short-term prediction filter which require 5; 5; 5; 4; 4; 3; 2; and 2 bits, respectively. Apart form this a new 5 bit quantizer of maximum amplitude of excitation pulse has been derived. The number of quantization levels for the remaining quantizers have been unchanged. However, the optimization of their characteristics has been carried out.

Indeed, the obtained reduction of the number of bits in the speech frame has caused the decrease by 2.6 dB of the average PQR, however, the quality of synthesized speech turned out to be sufficiently good. The analysis of the average signal-to-noise ratio [2] of the reconstructed speech for the test speech signals indicates only small decrease by 0,7 dB. However, taking into account the average log spectral distortion of a speech frame a small increase by 0,1 dB is observed.

The comparable audio tests of the standard and modified speech encoder for 2 women voices and 2 men voices have shown that subjective quality of modified speech encoder is fully acceptable as one can see in table 1.

Table 1. Results of the subjective quality of speech reconstruction. Notation: (+) – higher quality, (-) – lower quality, (=) – the same quality.

Evaluating person	Woman 1 voice		Woman 2 voice		Man 1 voice		Man 2 voice	
	GSM	PROJECT	GSM	PROJECT	GSM	PROJECT	GSM	PROJECT
D.G.	+	–	+	–	–	+	+	–
P.C.	–	+	=	=	=	=	=	=
R.S.	+	–	+	–	+	–	=	=
J.S.	=	=	=	=	=	=	=	=
M.B.	=	=	=	=	=	=	=	=
S.K.	–	+	–	+	+	–	–	+

The most important result obtained with respect to the modified speech encoder is the reduction of the speech date rate from 13 kb/s to 12,5 kb/s. Precisely speaking, the modified speech encoder is generating 250 bits instead of 260 bits per speech frame of the standard speech encoder.

In table 2 one can find the number of bits needed for the parameters of standard encoder and the modified one.

Table 2. The number of bits required for parameter coding of standard and modified encoder.

Parameters	Number of bits (GSM)	Number of bits (PROJECT)
Parameters of short term prediction filter	36	30
Initial time phase of excitation pulses	8	8
Maximum amplitude of excitation pulses	24	20
Amplitudes of remaining excitation pulses	156	156
Gain of long-term prediction filter	8	8
Lag of long-term prediction filter	28	28
Total	260	250

3. IMPROVED CHANNEL CODING

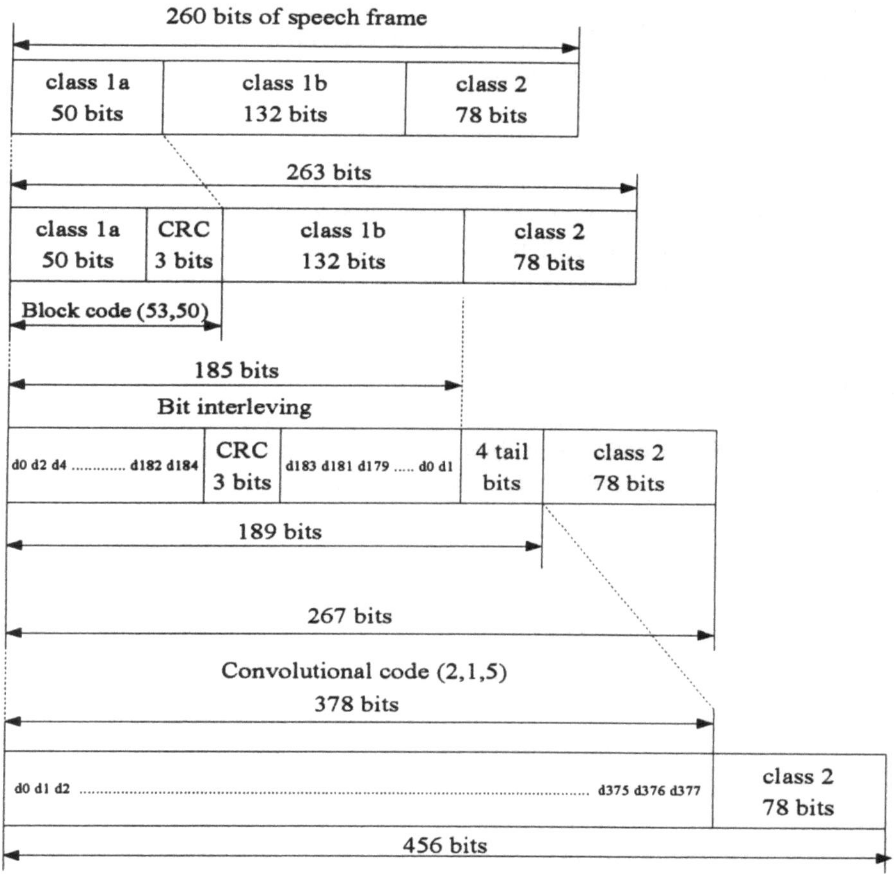

Figure 2. Standard channel coding of speech frame

The standard speech frame containing 260 bits is further undergoing channel coding. First, it is encompassing error detection coding of class 1a bits, by the use of CRC adding 3 parity bits. Next, to the obtained block together with the class 1b bits additional four 0 bits are added to reset the convolution coding employed as error correction coding. Further details of standard coding are shown in Fig 2.

The total number of bits in the speech frame reduced by 10 bits can enable to design a new channel coding with enhanced error protection and to improve the performance of reconstructed speech at the poor though still acceptable channel conditions.

Thus, instead of a simple error detection coding concerning the class 1a bits of standard coding a new error correction coding has been implemented. As a result the rate of discarded speech frames has been reduced and the performance of reconstructed speech has been improved. The code that has been used is BCH (63, 57). It has the error correction capability of 1 error and its generating polynomial is given by

$$g(u) = u^6 + u + 1 \quad (1)$$

To obtain some error detection capability an extended BCH code has been employed with one added parity bit. A new error protection coding is shown in Fig. 3. One can see in Fig.3 that 7 zero bits have extended the lenght of class 1a bits to carry out block error coding. Of course, on the channel only the sequence of 50 class 1a bits and 6 parity bits is transmitted. The new error coding of speech frame is shown in Fig. 4. As one can see the number of bits allocated to class 1b and class 2 has been slightly changed if compared with the standard channel coding. This change is motivated by convenient arrangement of BCH code parity bits and the format similar to the standard one. Thus, the total number of coded bits in the speech frame remains unchanged.

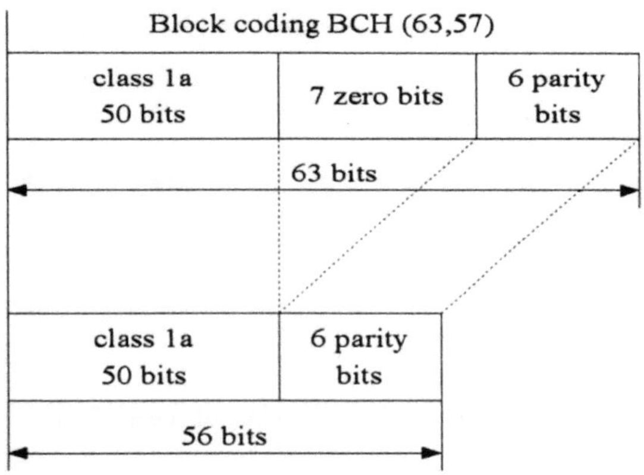

Figure 3. Error correction coding of class 1 a bits

4. AN ALGORITHM OF CHANNEL DECODING

In the standard decoding algorithm the received speech frame is discarded if an error is detected in the sequence of class 1a bits. A new channel coding requires some changes in the decoding algorithm (see [5]) that are shown in Fig.5.

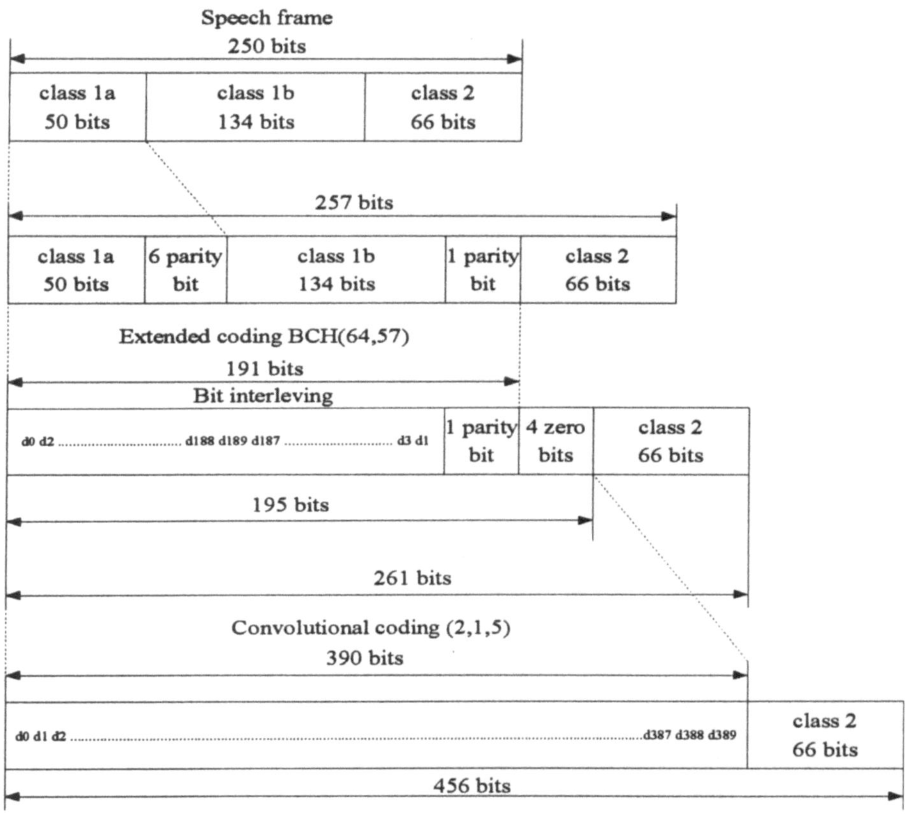

Figure 4. Channel coding of speech frame containing 250 bits

First of all, the syndrom of received class 1a bits must be calculated. This requires to add 7 zero bits (see Fig. 3). If there are no errors an acceptance decision is undertaken. If there are some errors, the speech frame is discarded provided that the error appears on the forbidden positions (7 positions), otherwise the error correction is carried out and the parity bit is calculated. In the case of identical values of parity bit calculated and received, the speech frame is accepted. In another instance it is discarded and the procedure of previous frame substitution is initiated.

The improved speech and channel coding/decoding has been tested through simulation in order to compare the results with that for the standard channel coding/decoding provided that signals are transmitted on the channel with fading, noise and at different mobile station speeds. The evaluation was based on the indication by the listeners, which one out of two reconstructed speech test signals with the use of standard and improved speech and channel decoder is better in quality. Some of the results (see [5]) for urban propagation environment are shown in tables 3 and 4.

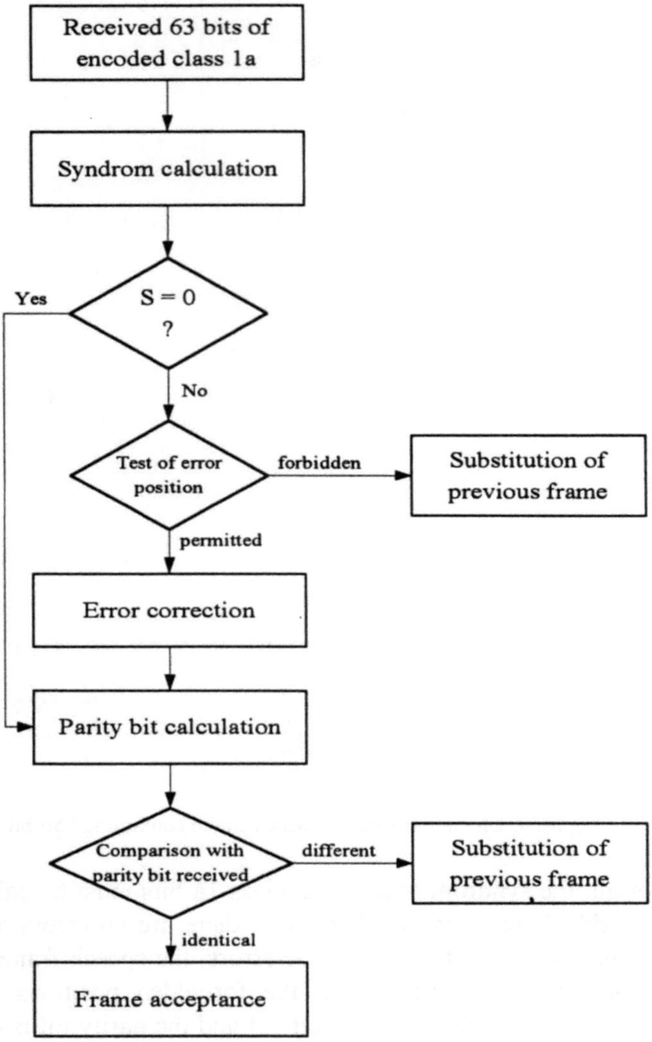

Figure 5. A decoding algorithm of speech frame for new channel coding

As one can see the improved speech and channel coding/decoding enables to obtain better performance particularly at the poor channel conditions i.e. at low E_b/N_0.

Table 3. Simulations results of the subjective quality of speech reconstruction of a woman voice for the standard and improved speech and channel coding/decoding. Propagation environment: TU 50. Notation: (+) – higher quality, (-) – lower quality, (=) – the same quality

Evaluating person	E_b/N_0 [dB]									
	4,5		5,5		6,5		8,5		10,5	
	GSM	PROJECT	GSM	PROJECT	GSM	PROJECT	GSM	PROJECT	GSM	PROJECT
R.S.	–	+	–	+	=	=	–	+	–	+
D.G.	–	+	–	+	–	+	–	+	–	+
J.S.	–	+	–	+	–	+	–	+	+	–
S.K.	–	+	–	+	–	+	–	+	–	+
P.C.	–	+	–	+	–	+	–	+	–	+

Table 4. Simulations results of the subjective quality of speech reconstruction of a man voice for the standard and improved speech and channel coding/decoding. Propagation environment: TU 50. Notation: (+) – higher quality, (-) – lower quality, (=) – the same quality

Evaluating person	E_b/N_0 [dB]									
	4,5		5,5		6,5		8,5		10,5	
	GSM	PROJECT	GSM	PROJECT	GSM	PROJECT	GSM	PROJECT	GSM	PROJECT
R.S.	–	+	–	+	–	+	–	+	+	–
D.G.	–	+	–	+	–	+	–	+	–	+
J.S.	–	+	–	+	=	=	=	=	=	=
S.K.	–	+	–	+	–	+	–	+	–	+
P.C.	–	+	–	+	+	–	–	+	–	+

5. CONCLUSIONS

The results obtained indicate that the improvement of performance of GSM system is possible particularly in the areas where poor channel conditions exist. This concerns the areas at the cell borders and the areas of propagation shadow. By improving the performance of reception the degree of coverage by the system can also be increased.

REFERENCES

[1] ETSI, *European digital cellular telecommunications system* (Phase 2), (Sophia Antipolis CEDEX - FRANCE), 1995.

[2] R.A. Salami, L.Hanzo, R.Steele, K.H.J.Wong, and I.Wassell, *Mobile Radio Communication*, pp. 186-346. Pentech Press, 1 st ed., 1992.

[3] J.Stachurski, *A Pitch Pulse Evolution Model for Linear Predictive Coding of Speech*, Ph.D. dissertation, McGill University, Montreal, Canada, 1997.

[4] D.Godyń, *A Choice of Characteristics of Nonlinear Quantizers for Spectrum Efficient Speech Coding in GSM System*, Proc. Nat. Comm. Conf. in Polish, KST'98, vol. A, pp. 179-185, Bydgoszcz, Poland, 1998.

[5] D.Godyń, *Investigation of Spectrum Efficient Techniques of Coding for GSM System*, Ph.D. dissertation, Technical University of Gdańsk, Gdansk, Poland, 1999.

A Picocellular CDMA/TDD Overlay on GSM

Piotr Kaczorek [*], Dominik Rutkowski [**]
[*] *Department of Marine Radio Electronics, Gdynia Maritime Academy, Poland*
[**] *Department of Radio Communication Systems, Technical University of Gdansk, Poland*
pik@wsm.gdynia.pl, nick@pg.gda.pl

Key words: GSM system , CDMA/TDD, UMTS/IMT2000

Abstract: To improve the utilisation of frequency bands assigned to GSM system, an idea of CDMA overlay seems to be interesting. In this paper a picocellular CDMA/TDD overlay on GSM system has been considered. It is corresponding to a TDD mode of WCDMA proposed by ETSI, ARIB and TIA for the future third generation UMTS/IMT2000 system. The overlay is proposed for indoor operation in the area of macrocell of GSM system. As it turns out the implementation of overlay can significantly increase the spectrum efficiency in the frequency band that is shared by GSM system and the overlay. The picocellular CDMA/TDD overlay can be used both for speech and data transmission.

1. INTRODUCTION

The frequency band assigned to each radio communication system is a valuable resource and as time goes by we will observe more frequently the coexistence of autonomous radio communication systems sharing the same frequency band with the constrained level of mutual interference.

In general, by CDMA overlay we understand a mobile communication system based on DS-CDMA technique that is sharing frequency band with another mobile communication system based on TDMA or FDMA technique. In the literature one can find many publications in this topics [5-9]. However, only macrocellular overlay, covering fully the area of GSM cells, has been considered so far. The paper [10] analyses the CDMA overlay in an indoor picocell that is located in the area of a GSM macrocell. Such an overlay has the potential of enhancing the overall spectrum efficiency and is able to provide additional capacity in the areas of excessively large traffic, e.g. business districts and commercial centres. Clearly, the condition that must be met to use an overlay is the simplicity of implementation. This is why in [10] an overlay based on WCDMA standard, worked out by ETSI for UMTS/IMT-2000 system, has been proposed. Of course, CDMA overlay should be designed in such a way as to avoid excessive interference to GSM system and it should not restrict in any form GSM system's functioning.

Thus, the chance of using the overlay and its capacity depends upon the mutual interference between GSM system and the overlay. The analysis of such interference in case of CDMA/FDD overlay proposed in [10] has shown that important factor limiting a location and capacity of picocellular overlay is the interference experienced in the uplink of CDMA/FDD overlay and coming from GSM system's mobile terminals.

In the following, a problem of designing a picocellular CDMA overlay on GSM system will be considered provided that WCDMA/TDD standard is employed. The TDD mode enables to increase the range of possible locations of a picocell and to make its capacity independent on location.

2. CDMA OVERLAY

Let us assume that frequency band of *12 MHz* is assigned to one GSM system operator, the cluster of cells is *4* and the base stations use three-sector antennas. For a uniform traffic in the whole service area each sector of a cell has $N_T = 5$ frequency channels to its disposal.

The frequency channel of CDMA overlay, having the bandwidth of *4.8 MHz*, consists of *24* consecutive GSM system's frequency channels in the uplink. Thus, it has two channels, on the average, that are assigned to each sector of GSM system as it is shown in Fig. 1.

Known solutions of macrocellular CDMA overlay consist in narrowband notch filters in the transmitters and receivers of overlay. The characteristics of notch filters used in a given sector of overlay must be chosen adequatly to attenuate frequencies utilised in that sector by GSM system [5-7]. This approach can also be used in the picocellular overlay considered, however, only in such a case when a picocell is located somewhere in the middle of a sector of GSM system's macrocell (see Fig. 1.A). If a picocell is located close to the boundary between sectors, additional filters must be used due to powerful interference coming from the overlay and affecting GSM terminals in the neighbour sector (see Fig. 1.B). These additional filters should attenuate frequencies used in the nearest sector of GSM system. Thus, their characteristics should depend upon specific location of overlay.

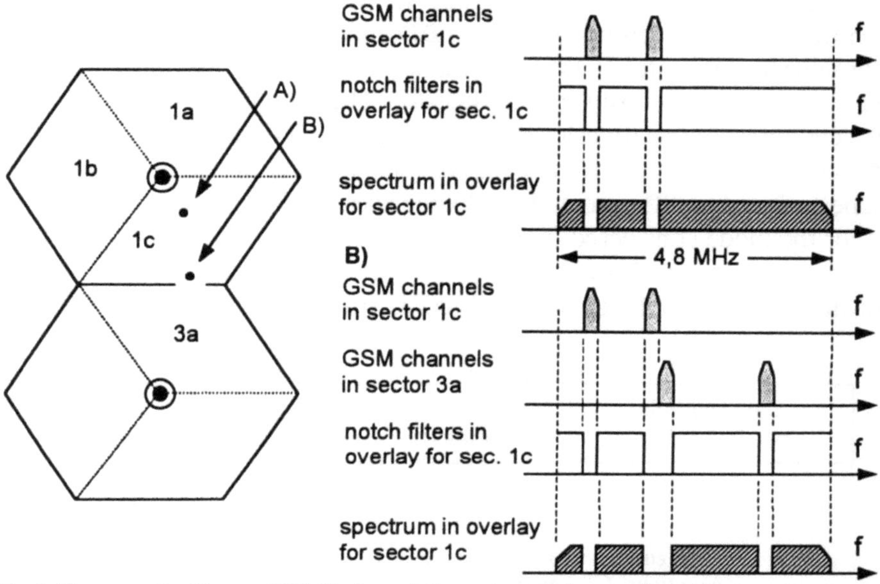

Fig.1. The way of setting up CDMA channel for a picocellular overlay: (A) picocell is located inside a sector of GSM system (B) picocell is located close to sector boundary.

In the following, we are considering the overlay in which filtering is encompassing frequencies used by two GSM sectors, including that with the overlay. It means that filtering will concern at most 4 GSM channels.

3. TDD MODE IN UMTS/IMT-2000 SYSTEM

The ETSI proposal of WCDMA standard for UMTS/IMT-2000 system is foreseeing in TDD mode the use of mixed TDMA/CDMA multiaccess scheme. Some of the parameters and characteristics for that mode are given in Table 1.

Table 1. Parameters and characteristics of WCDMA/TDD [1]

Multiaccess	TDMA/CDMA
TDMA frame	10 ms
Number of timeslots in TDMA frame	16
Chip rate	4,096 Mcps
Spreading rate	16
Channel data rate	128 ks/s (256kb/s)
Modulation	QPSK

As one can see, a 10 ms TDMA frame is subdivided onto 16 timeslots. Each 625 μs timeslot can be assigned for the uplink or downlink. This provides for the flexibility in using TDD mode for different services and environments. Due to CDMA technique more than one burst may be transmitted simultaneously in one timeslot. The orthogonal transmission of different bursts in one timeslot is assured by the use of orthogonal spreading sequences. Of course, different bursts (equivalently spreading sequences) can be assigned to different users or to one user.

For the mode considered two types of bursts have been defined. Further, we are going to analyse TDD mode with type 1 burst which can be transmitted in both uplink and downlink. Its structure is shown in Fig. 2.

Data symbols 61 (976 chips)	Midamble 512 chips	Data symbols 61 (976 chips)	GP 96 CP

← 625 μs →

Figure 2. Burst structure of the traffic burst 1 [1].
Notations: GP - guard period with 96 chip periods (CP).

All bursts in all timeslots are spread with the same spreading factor of $G = 16$. One can get different transmission rates by assigning different number of timeslots and/or different number of bursts/timeslot for a particular user. Example of mapping 64 kbps data service is given in Fig. 3.

The results of simulation described by ARIB [2] indicate that for indoor 64 kbps LCD (*Long Constrained Delay*) service the required $(E_b/N_0)_{min} = 1.2\ dB$ and $(E_b/N_0)_{min} = 5\ dB$ for the downlink and uplink, respectively.

One can prove that maximum number of bursts that may be sent in one timeslot is $N_{CB}^{(dn)} = 13$ for the downlink and $N_{CB}^{(up)} = 6$ for the uplink, respectively. Through the adequate asymmetric assignment of timeslots between downlink and uplink one can achieve $N_C = 10$ duplex channels 64 kbps each.

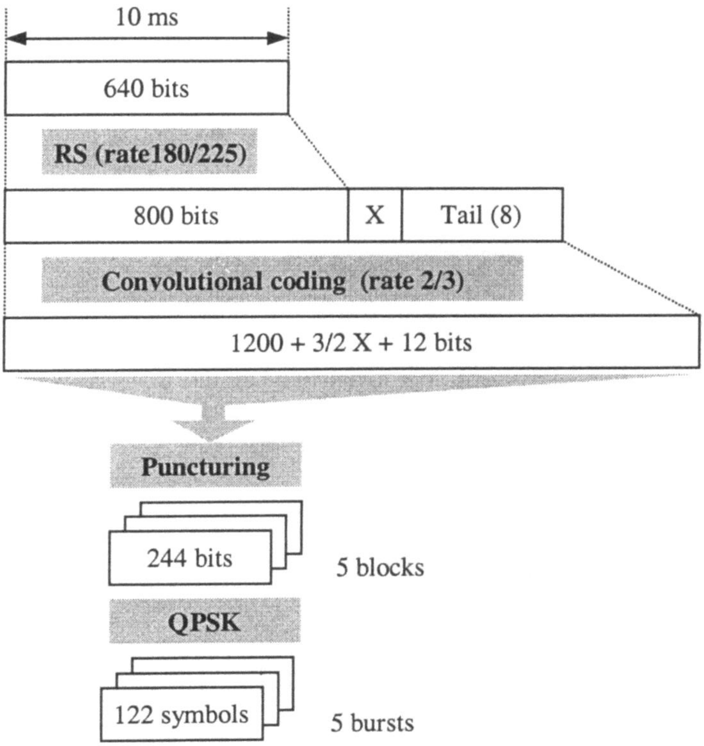

Figure 3. Channel coding for 64 kbps data service [1].
Notations: RS - Reed-Solomon coding, X - signalling bits

4. INTERFERENCE IN CDMA OVERLAY

The analysis of interference coming from GSM system and affecting CDMA overlay has been carried out for the uplink of CDMA overlay. The way of analysis is similar to the one shown in [10]. The propagation model adopted assumes the path loss proportional to the fourth power of distance. Besides, fading, shadowing, and thermal noise are omitted and ideal power control is employed.

The way of calculation the interference power coming from GSM system is illustrated in Fig. 4. The location of picocell with respect to the base station BS_0 of GSM system is described by angle φ_p and distance r_p. The source of interference is a base station BS_j of GSM system that is serving mobile terminal MS_j. The location of MS_j with respect to BS_j is given by angle γ_j and distance r_j.

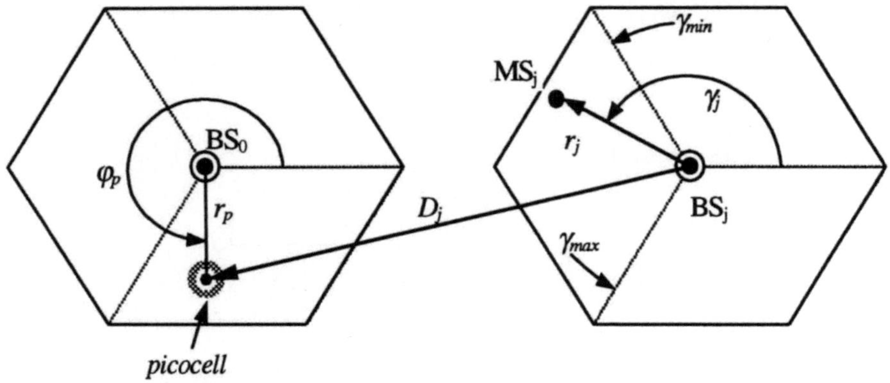

Figure 4. The way of calculation the interference coming from GSM system in the uplink of CDMA overlay.

The average interference power coming from BS_j that is being received in CDMA overlay depends on the location of picocell. At the assumptions made one can find it as

$$\overline{I_j(r_p,\varphi_p)}=\frac{1}{A_W}\frac{1}{\frac{\sqrt{3}}{2}R^2}\int_{\gamma_{\min}}^{\gamma_{\max}}\int_{0}^{R_{\max}(\gamma)}\left(\frac{r_j}{D_j}\right)^{\alpha}S_T\cdot r_j dr_j\, d\gamma_j \qquad (1)$$

where S_T is the nominal power of signal transmitted by MS_j and received by BS_j, D_j is a distance between picocell and MS_j. The angles γ_{min} and γ_{max} delimit the sector considered and $R_{max}(\gamma)$ is a distance between BS_j and sector boundary. With the assumption made that power control is ideal one, the signal power received by all mobile terminals of GSM system is the same and equal to S_T. The symbol A_w in the expression (1) represents the building penetration loss. In Table 2 typical values of building penetration loss are given. In our considerations more critical value of $A_w = 10\ dB$ has been chosen.

Table 2. Building penetration loss [11].

Environment	Penetration loss (dB)
Urban, Suburban	15
Residential, Rural	10

One can find the total interference power by summing up the average power for each of the sectors

$$I_T(r_p,\varphi_p) = N_{T/C} \sum_j \overline{I_j(r_p,\varphi_p)} \tag{2}$$

where $N_{T/C} = 2$ results from the conclusion made before that the overlay radio channel of overlay is including on the average two radio channels of each GSM sector (see Fig. 1).

The total interference power is upper-bounded by

$$I_T(r_p,\varphi_p) \leq I_{T\max} = 0{,}032\, S_T \tag{3}$$

Thus, the carrier-to-interference ratio *CIR* in the uplink of CDMA overlay is given by

$$CIR_C^{(up)} = \frac{GS_C}{(N_{CB}^{(up)} - 1)S_C + I_T} \tag{4}$$

where N_{CB} is the number of overlay bursts/timeslot, G stands for processing gain and the relationship between nominal power in overlay and GSM system, $S_C = 0.36\, S_T$, is calculated similarly to [10].

Since the analysis doesn't take fading, shadowing and thermal noise into account, then we can't directly use *CIR* for the performance evaluation of overlay. Instead, we can find the degree of degradation of *CIR* in overlay affected by interference coming from GSM system

$$\delta CIR_C^{(up)} = \frac{CIR_C^{(up)}{}_{(I_T=0)}}{CIR_C^{(up)}{}_{(I_T)}} \tag{5}$$

Taking (3) and (4) into account, we obtain

$$\delta CIR_C^{(up)} \leq 0.06\, dB \tag{6}$$

We can draw the conclusion that the influence of GSM system on the overlay is negligible. One can get similar results for the downlink of overlay. We can see that CDMA overlay attains in practice the same capacity as WCDMA system does independently on the location of picocell with respect to the sector of GSM system's macrocell.

5. INTERFERENCE RANGE OF CDMA OVERLAY

As we know, in CDMA overlay considered a TDD mode of transmission is used. Thus, the interference coming from overlay affects only downlink of GSM system. In the following, we assume that a picocell with CDMA overlay has a shape of a circle with a radius of R_p and the base station is located in the centre. A mobile terminal of GSM system is located at a distance D_j from the picocell. This is illustrated in Fig. 5.

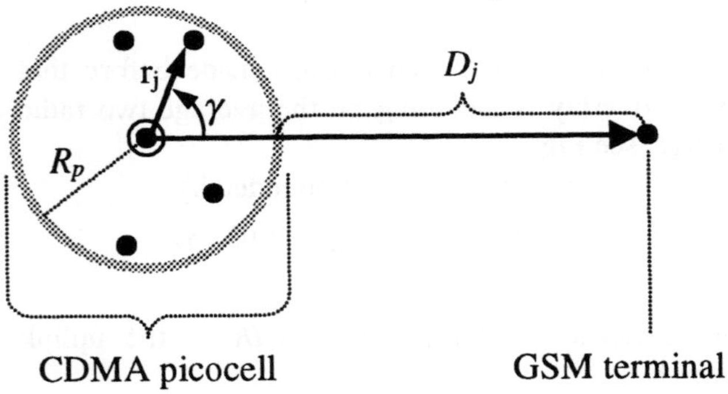

Figure 5. The way of interference range calculation of CDMA overlay.

In view of the assumptions made, the average interference power arriving at the GSM mobile terminal from the base station of overlay can be found from the expression

$$\overline{I_C} = \frac{N_{CB}}{HA_W} \frac{1}{\pi R_p^2} \int_0^{2\pi} \int_0^{R_p} \left(\frac{r_j}{R_p + D_j} \right)^4 S_C \cdot r_j dr_j \, d\gamma \tag{7}$$

After calculations we get

$$\overline{I_C} = \frac{1}{3} \cdot \frac{N_{CB}}{HA_W} \left(\frac{R_p}{R_p + D_j} \right)^4 S_C \tag{8}$$

where S_C is the nominal power of signal received by overlay terminals, $N_{CB}^{(dn)}$ is the number of bursts/spreading sequences transmitted in one timeslot, and A_w stands for the building penetration loss.

Due to narrower channel bandwidth in GSM system than in overlay, only a fraction of interference power is affecting GSM terminal. This is reflected in (8) by the interference reduction factor H given by

$$H = \frac{W_C}{W_T} = 24 \tag{9}$$

where W_C and W_T is the channel bandwidth of CDMA overlay and GSM system, respectively.

Now, we can find the carrier-to-interference ratio CIR_T in GSM system in the presence of interference coming from CDMA overlay as

$$CIR_T = \frac{S_T}{\overline{I_T} + \overline{I_C}} \tag{10}$$

The average interference power I_T in the downlink of GSM system can be calculated in the same way as in [9]. As a result we obtain

$$\overline{I_T} \approx 0{,}0028\, S_T \tag{11}$$

Substituting (8) and (11) into (10), we have

$$CIR_T = \frac{S_T}{0{,}0028\, S_T + \frac{1}{3}\frac{N_{CB}}{HA_w}\left(1 + \frac{D}{R_p}\right)^{-4} S_C} \tag{12}$$

For the same reasons as before we can't use directly formula (12) to the evaluation of interference range for CDMA overlay. Instead, we can find the degree of degradation of CIR_T in the GSM mobile terminal caused by interference coming from overlay. As a reference level the value of CIR at the full load of GSM system and lack of interference coming from overlay has been adopted, i.e.

$$\delta CIR_T = \frac{CIR_{T(N_{CB}=0)}}{CIR_{T(N_{CB},D_j)}} \tag{13}$$

The results of calculations are shown in Fig. 6. As one can notice the interference range of CDMA overlay is small and doesn't exceed two radii of picocell for $\delta CIR_T = 0.1\ dB$.

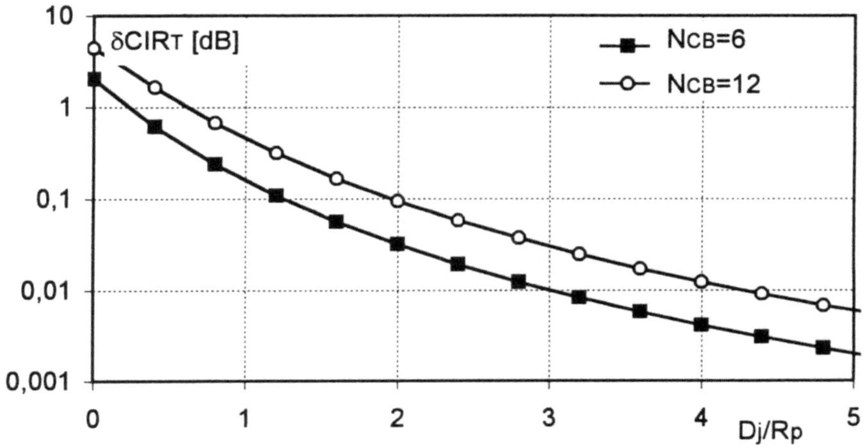

Figure 6. Degradation of CIR vs. D_j/R_p in GSM system due to interference coming from CDMA overlay. Notations: D_j - distance between GMS mobile terminal and overlay, R_p - radius of picocell, N_{CB} - number of bursts/timeslot.

Having obtained interference range of CDMA overlay one can get the region of permitted picocell locations. In Fig. 7. the boundaries of the regions are shown in which CDMA overlay location is possible in the most unfavourable situation when the radius R_p of picocell is 10 times smaller than the radius R_m of GSM macrocell. Different boundary lines concern different number of filters employed. When filtering in overlay is encompassing only the frequencies of own GSM sector a relatively small area of picocell locations is possible (some *36%* of sector area - see Fig. 7.A).

However, when filtering is including the frequencies of two GSM sectors the allowed area of picocell locations is significantly increased (some *84%* of sector area - see Fig. 7.B). Further increase of the number of filters isn't reasonable since the benefit is small and some degradation of transmission quality in overlay occurs.

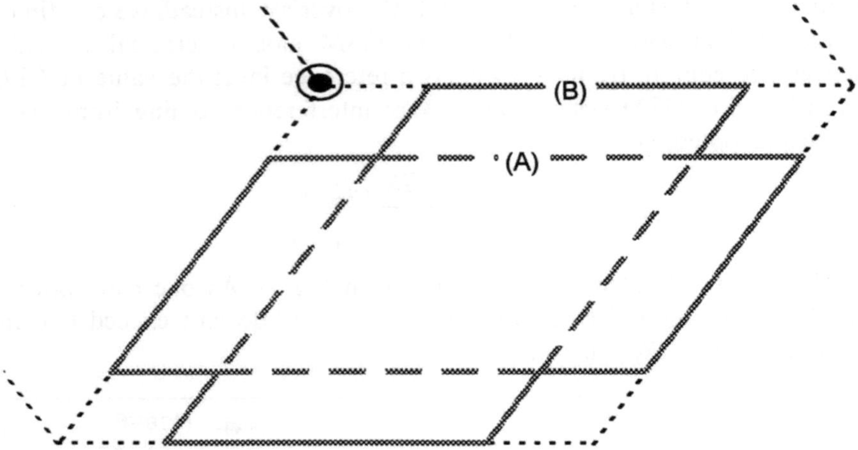

Figure 7. The region of permitted picocell locations.
(A) filtering in overlay of the frequencies of own GSM sector
(B) filtering in overlay of the frequencies of two GSM sectors including the own one.

6. CONCLUSIONS

The results obtained suggest that a proposed CDMA overlay can significantly increase the spectrum efficiency of the frequency band shared by GSM system and the CDMA overlay and can provide additional capacity for services.

One can also prove that larger percentage of GSM service area can be used for picocell CDMA overlay with TDD mode than with FDD mode. Apart from this, the implementation of narrowband filters is easier with TDD particularly in the receivers of overlay.

It is also worth to mention that known plans of the evolution and implementation of third generation systems forecast the integration of WCDMA system with GSM system [1-4]. Thus, the proposed overlay can be an attractive platform for the coexistence of GSM and WCDMA system.

Practical implementation of the proposed CDMA overlay depends on the correct functioning of WCDMA system despite of the filters employed. The results of simulation [5] and hardware tests [8] indicate that this is realisable.

REFERENCES

[1] ETSI: „UMTS Terrestrial Radio Access (UTRA); Concept evaluation", Dec. 1997.

[2] ARIB: „Japan's Proposal for Candidate Radio Transmission Technology on IMT-2000: W-CDMA", June 1998

[3] R.Prasad, T.Ojanpera „An Overview of CDMA Evolution Toward Wideband CDMA" *IEEE Comm. Surveys*, No 4, 1998

[4] J.Schwarz da Silva, B.Barani, B.Arroyo-Fernan-dez „European Mobile Communications on the Move" *IEEE Comm.* Vol.10, 655-668, Feb.1997

[5] D.M.Gricco, D.L.Shilling „The Capacity of Broadband CDMA Overlaying a GSM Cellular System" Proc. IEEE VTC, 31-35, Stockholm, June 1994

[6] D.L.Schilling et al. „Broadband CDMA overlay" International Journal of Wireless Information Networks, Vol.2, No 4, 197-221, Sept..1995

[7] P.Koorevaar, J.Ruprecht „Frequency Overlay of GSM and Cellular B-CDMA" *Proc. IEEE VTC.*, 497-501, Stocholm, June 1996

[8] L.B.Milstein, D.L.Schilling „The CDMA Overlay concept" *Proc. IEEE VTC*, 476-480, Stockholm, June 1996

[9] P.Kaczorek, D.Rutkowski „A Comparison of Narrowband and Broadband Overlay of Cellular CDMA on GSM", *paper accepted for presentation on IEEE VTC2000-Spring*, Tokyo 2000

[10] P.Kaczorek, D.Rutkowski „A picocell CDMA Overlay on GSM system for Fast Data Transmission" *paper accepted for presentation on Nat. Radio Comm. Conf. KKRR'2000*, Poznań, Poland, June 2000

[11] A.Mehrotra „Cellular Radio Performance Engineering" Artech House, 1994

Design of Interoperability Checking Sequences Against WAP

O. Kone and J-P. Thomesse
LORIA - INPL, Campus Scientique B.P. 239 Vandoeuvre-les-Nancy 54506 France.
kone@loria.fr, thomesse@loria.fr

Key words: WAP, Application and Service Design, Design of Interoperability Tests.

Abstract: The deployment of a telecommunication service involves technical issues such as cellular transmission, routing etc, which are transparent to the user. Electronic mail or Short Message System are good examples of communication services used by persons provided with wireless terminals. The incoming generation of mobile equipment will access internet-like services and WAP (Wireless Application Protocol) is a standard destined to specify the related new architectures. WAP forum efforts will also concentrate on the development of tests aimed to insure the conformance and the introperability of WAP implementations. The contribution of this paper is the design of checking sequences for the interoperability of WAP implementations. Our work mainly contributes to phase two of WAP conformance process review.

1. INTRODUCTION

Many research works have been taken in various fields of wireless communication to solve technical issues such as cellular transmission and routing [9, 18, 14], application and services [8]. The dynamism in wireless communication is due to the permanent evolution of mobile technology. For instance, the new generation of mobile equipment will not be restricted to classical telephony applications. More than standard communication services, internet-like services will be also accessible from mobiles.

"... The world's largest consumer payments organization and the world's number one mobile phone manufacturer have signed a global agreement to develop ways in which financial institutions and mobile

phone operators can offer secure payment services to their customers via a mobile phone. The organizations will carry out joint market development activities and pilot technical payment alternatives ..."

WAP forum News, San Francisco, 8 February 2000.

The previous annoucement shows that the availability of Internet-like services from mobile terminals is becoming a reality. WAP (Wireless Application Protocol[16]) is a standard destinated to specify the concerned new architectures. The design of WAP systems will follow the classical protocol development cycle including *specification, implementation* and *testing.* A wide range of implementations belonging to different manufacturers will be available on the market. These implementations will face problems such as reliability, conformance/certification and interoperability (due to implementations heterogeneity).

Conformance testing consists of checking wether a given implementation meets the requirements of the standard specification [7, 11]. *Interoperability testing* consists of checking wether different implementations successfully communicate and interwork according to the standard. WAP forum has started to define a testing framework which will be followed in two steps. The first step is concerned with conformance testing of WAP application part and will issue certification labels for WAP products. The second step is concerned with the interoperability of WAP layer-to-layer products and is also referred to as *compliance testing.*

The contribution of this paper is the design of checking sequences for the interoperability of WAP implementations. The proposed sequences are test pattern which can show that different client and server entities can interoperate or not.

In the following section, we recall the main features of WAP architecture, and we introduce the methodology we adopted for testing WAP. Section 3 describes our experiment in modelling and selection of test patterns for WAP.

2. A METHODOLOGY TO TEST WAP INTEROPERABILITY

2.1. ARCHITECTURE OF WAP NETWORKS AND SERVICES

WAP is aimed at enabling mobile terminals to access Internet-like services. Within the WWW (World Wide Web), a computer program can ask another one to execute some request. The first program is called *client* and the second one is called *server.* In order to insure interoperability, the WWW is built over standards such as HTTP [1]

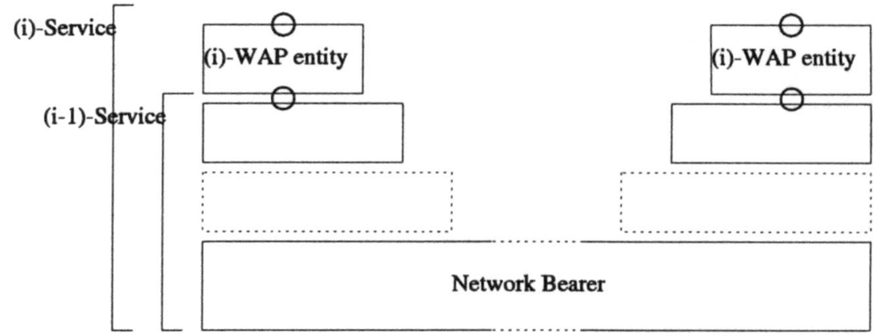

Figure 1 Structure of WAP services

and HTML [2]. The WWW model has influenced the WAP model very much. Distributed wireless applications are also composed of clients and servers. Mobile clients will have access to both WAP-origin and WWW servers. A WAP-origin server is, for example, a Wireless Telephony Application server which is a direct access point of a given wireless network provider to the WAP-client. But the communication with a WWW server requires to go through a WAP-proxy server which stands for an intermediary node between the two technologies. The mobile application will be programmed with WML (Wireless Markup Language) which is very close to HTML. The mobile terminal will be provided with a microbrowser for WML. WAP functionnalities will be carried out by a family of protocols that constitute the WAP-stack.

The structure of WAP model (figure 1) resembles the one of OSI basic reference model. An (i)-WAP layer is composed of (i)-WAP entities using a (i-1)-WAP service. One of the particularities of WAP structure is that some (i)-service can be directly used by upper layers or applications. For example, the Transaction layer or further defined applications can directly access the Transport layer.

Normative references tell more about WAP. The reader may report to the standards for detailed information.

2.2. A FRAMEWORK TO DISTRIBUTED/INTEROPERABILITY TESTING

This section describes the main characteristics of the methodology adopted for WAP tests development. The so-called TOP *methodol-*

ogy [10, 12, 13] has been a contribution to normative testing area for the interoperability of communicating systems. Most experiments with TOP have been taken with standards such as ATM Adaptation Layer, ATM-ABR or OSI protocols [3, 4].

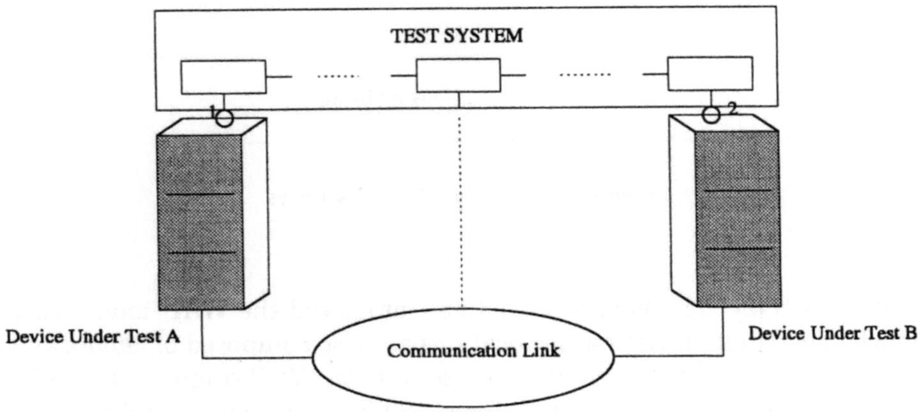

Figure 2 **Interoperability Testing Architecture**

Figure 2 depicts an abstract view of the TOP test architecture. The Service Acces Points (SAP) are labeled with 1 and 2. The test system interacts with the devices under test through these SAP. An optional interaction point enables it to access the PDU (Protocol Data Unit) at lower levels.

Behavioural testing. The test campaign consists of submitting interactions to implementations while observing their reaction. Test sequences (or checking sequences) can be automatically computed if formal models of reference specifications are available. Our approach works with input/output automata for modelling the specifications and the test purposes ($\stackrel{\text{def}}{=}$ *the properties to be checked*). The algorithm implemented in the TOP test generation tool is an on the fly seacrh [12] which explores the specifications of the devices under test and computes pertinent test pattern which can be used to experiment the implementations. More details on the TOP methodology and its applications are available in [10, 13].

Using TOP against WAP. Compare figures 1 and 2. As explained before, each WAP layer is intended for providing some service. This service is to be used by actual or further upper layers and applications. Obviously, the reliability of a given layer depends on the one of the underlying services. So checking the correctness of the service provided by implementations is fundamantal. The WAP architecture enables access to each layer of the WAP stack and the TOP architecture is powerful to check a distributed system through its service boundary. In practice, the connection to the service access points of the devices may require a few intrumentation. Typically, an interface may be used to encode/decode the interactions (or signals) exchanged between the test system and the devices under test. In figure 2, the white part of the devices represents such possible interfaces, while the grey part represent their internal design.

Moreover, WAP specifications include, for each layer, the protocol description by state tables. We modeled the specifications (client entity, server entity) with finite input/output automata which were further input to our test generation tool. The next section tells more about tests design for WAP.

3. EXPERIMENT WITH THE DESIGN OF CHECKING SEQUENCES

Because of the recursive structure of WAP service layers, ((i)-Service ≡ (i)-Protocole + (i-1)-Service), our work can be applied to each WAP layer. But in this paper, we will focus on our experiment with the WAP session layer (WSP), which is the top of the layers underlying the WAP Application Environment.

3.1. MODELLING ISSUES

WSP enables client and server applications to exchange contents through a maintained communication session. WSP specifications [17] include a connectionless protocol over a datagram service, and a connection-mode protocol over a transaction service. In the sequel, we consider the connection-mode protocol only, as the other one is fairly simple.

WSP is mainly featured to establish and release a session between a client and a server, exchange contents of the two applications, suspend and resume the session.

The WSP is defined over a set of service primitives and their ordering by time sequence charts. A service primitive is of the form `S-Service-type`, (e.g. `S-Connect-req`), where `S` is the label of the session layer, `Service` indicates the name of the service (e.g. `Connect`). The type (e.g. `request`) is also indicated. Protocol operations are performed through the use of a transaction (`TR`) service. The PDU exchanged between the WSP peer-entities are encapsulated in the `TR` primitives: `TR-Service-PDU` denotes the encapsulation of a given `PDU` in a `TR-Service` primitive. For example, `TR-Invoke-Connect` represents the encapsulation of the `Connect` PDU in `TR-Invoke`. The session protocol is described with a set of state tables. We modeled the client and the server protocol entities with Input Output automata as this model was the formalism used by our test generation tool. In order to check the correctness of our model we translated it in SDL language [5] so that it could be verified with the `ObjectGeode` simulator/debugger. The specification file contained about 2500 lines of SDL code. Of course, the verification of the model revealed some errors of modelling (some transitions were forgotten) and mainly some unused transitions. It appeared that the unused transitions corresponded to the reaction of entities face to the possible errors introduced by the underlying layer. In fact, for the time being, our test design methodology does not deal with error generation which falls in the category of robustness testing. Error cases due to the communication environment are various and not controllable. Our test design process is based on automatic computation of the behaviour part assuming a normal operating environment. Further studies will investigate error generation. Another specification issue concerns optional services (e.g the `Push` service). Optional services are not systematically included in our reference model. We described these services as SDL processes which may be added or not to the mandatory part of the specification.

3.2. TEST/CHECKING SEQUENCES

In the WSP specifications, implementation features are organised in groups of functionality (*Session creation, Session suspend/resume, Push facilities* etc). We used the guide of WAP Implementation Conformance Statement [15] as a basis to test selection. Even if test patterns are produced automatically with our tool, test purposes are to be defined formally before, and this work is done by hands. In this paper, we will not include all of the test purposes considered. We will present two examples to illustrate the experiment with WSP layer : The session creation feature and the connection redirection.

Session Creation. *Connecting the mobile terminal*

Session creation starts with an `S-Connect-req` (Connection Request), and ends with an `S-Connect-cnf` (Connection confirmation). The session creation abstract behaviour can be defined with the test purpose below described with a finite automaton. The syntax expresses one transition per line. Each line describes an interaction as well as the implementation which executes it. For example, "1" represents the identifier of the WAP-client implemenpation (first and last lines), "2" represents the identifier of the WAP-server implemenpation (second and third lines).

```
TP1 ? 1.S-Connect-req 1 TP2
TP2 ! 2.S-Connect-ind 2 TP3
TP3 ? 2.S-Connect-res 2 TP4
TP4 ! 1.S-Connect-cnf 1 TP5
```

Our test generation tool computes the client and server WAP-specifications according to the test purpose. The tool produces a test pattern which can be used to experiment the expected feature (in this example, the connection establishment).

The lines of the form `<XXX XXX XXX>` indicate the global state which may be actually reached by the distributed system: In the first line example, `NULL` represents the starting state of the client, the second `NULL` represents the starting state of the server, and `TP1` is the beginning of the behaviour to be checked. Between global states are displayed the interaction to be executed.

```
    <NULL NULL TP1>
1:   ? 1.S-Connect-req 1
   <NUCIA1 NULL TP2>
2:   ! TR-Invoke-Connect 1
   <CONNECTING NUCIA1 TP2>
3:   ! TR-Invoke-ack 2
   <CONNECTING NUCIA2 TP2>
4:   ! 2.S-Connect-ind 2
   <CONNECTING CONNECTING TP3>
5:   ? 2.S-Connect-res 2
   <CONNECTING CIC2A1 TP4>
6:   ! TR-Result-ConnectReply 2
   <CICEA1 CONNECTING-2 TP4>
7:   ! TR-Result-ack 1
   <CICEA2 CONNECTED TP4>
8:   ! 1.S-Connect-cnf 1
   <CONNECTED CONNECTED TP5>
```

1: A Connection request is received by implementation `1` (the client session entity).

2: The client invokes the transaction layer for sending its `Connect` PDU.

3: Upon reception of that PDU, the server (session entity) acknowledges it through the transaction layer.

4: The server sends a Connection indication to the upper layers.

5: Then it receives the response notifying that the application accepted the connection.

6: This notification is sent to the client, as a result of its request.

7: The client acknowledges the result.

8: The client session user is sent a confirmation indicating that the connection has been established.

Connection Redirection. *Connection redirected by WAP server*

```
   <NULL NULL TP1>
1:   ? 1.S-Connect-req 1
   <NUCIA1 NULL TP2>
2:   ! TR-Invoke-Connect 1
   <CONNECTING NUCIA1 TP2>
3:   ! TR-Invoke-ack 2
   <CONNECTING NUCIA2 TP2>
4:   ! 2.S-Connect-ind 2
   <CONNECTING CONNECTING TP3>
5:   ? 2.S-Disconnect-req 2
   <CONNECTING CITIA1 TP4>
6:   ! TR-Result-Redirect 2
   <CINUD1 CITIA21 TP4>
7:   ! 2.S-Disconnect-ind-USERREQ 2
   <CINUD1 TERMINATING TP4>
8:   ! TR-Result-ack 1
   <CINUD2 NULL TP4>
9:   ! 1.S-Disconnect-ind-Redirectparameters 1
   <NULL NULL TP5>
```

If a server can not respond (service unavailable or service moved), the connection may be redirected. In such case, the server session layer refuses the connection (line `5:`) and a `Redirect` PDU is sent back to the client (line `6:`) which may further try another connection.

The full execution of such test scenario by WAP client and server entities show their aptitude to interoperate. The occurrence of an unexpected interaction reveals an error.

4. CONCLUSION

Manufacturers and service providers pay high for tests development. Because of the competition involved in the industry of wireless communication, the reliability of products (and their correctness according to standards) has become a design criterion of first choice. Testing is a well known means of satisfying such criterion. But the quality of a test pattern also depends on the quality of the test production process. The approach used in this paper is based on automatic tests computation, which is a good way to insure tests soundness and test reproductibility (New standards are always emerging, so one can not imagine to design test by hands all the time). Our work is a contribution to the development of WAP standard destinated to new wireless communication systems. For space reasons we cannot include all the test patterns designed for WAP. All the mandatory functionalities have been experimented in dynamic testing. But we did not complete optional aspects. For instance, we considered the `Method Invocation facility`, but not for all the methods. On the one hand, most of these methods are optional. On the other hand, test design for methods is a repetitive work as it follows the same scheme. Only the reference of the method changes. As WAP is concerned with a very new technology, the related specifications are subject to modifications and this may influence the test patterns defined somehow. The test patterns will be stable when the standard will be definitively adopted and widely implemented during the incoming months.

References

[1] Hypertext Transfer Protocol. HTTP/1.1, RFC2068. R.Fielding et al. January, 1997.

[2] HTML 4.0 Specification, W3C Recommendation REC-HTML40-971218. D.Raggett et al. September 1997.

[3] ITU-T Recommandation Q.2110: B-ISDN ATM Adaptation Layer - ITU Telecommunication Standard Sector 1994.

[4] Information Processing Systems. Open Systems Interconnection. Transport protocol International Standard ISO 8073, 1988 (F)-AFNOR.

[5] ITU-T Recommandation Z.100: Specification and Description Language SDL, Contribution Com X-R215-E, 1987.

[6] ISO/TC 97/SC 16/WG1 IS 7498. Basic Reference Model for Open System Interconnection. 1983.

[7] ISO/IEC 9646, Information Technology - Open Systems Interconnection - Conformance testing methodology and framework-Part 1-5.

[8] Veikko Hara, Jarmo Harju, Jouni Ikonen, Jari Porras. Application of Distributed Workstation Environment to the Parallel Simulation of Mobile Networks IFIP TC 6 - Task Group Wireless Communications. 2nd PWC workshop Frankfurt am Main, Germany, December 1996

[9] Juntong Liu and Gerald Q. Maguire Jr. GMRM: An Efficient Routing Model for an Integrated Wireless Mobile Packet Switch Network. The 3rd Workshop on Personal Wireless Communication (PWC98), Tokyo, Japan, 1998.

[10] O.Koné, R.Castanet. Test generation for interworking systems Computer Communications, Elsevier Science Publishers Vol. 23 N.7 Mars 2000. pp 642-652.

[11] O.Koné, R.Castanet. Including Physical Time Constraints into Conformance. Proc. IFIP symposium on Protocol Specification, Testing and Verification. Warsaw, Poland, June 1995.

[12] O.Koné, R.Castanet, P.Laurençot. On the fly test generation for real time protocols. Proc. International Conference on Computer Communication and Networks. Louisianne, USA. October 1998.

[13] O.Koné. The TOP Methodology for the Interoperability of communicating systems. Technical Report INPL, Nancy December 1999.

[14] Tsuyoshi Tamaki, Hitoshi Aida, Tadao Saito. A Reserved Channel Scheme of Dynamic Channel Assignment for Multimedia Mobile Communication s Systems. IFIP TC 6 - Task Group Wireless Communications 2nd PWC workshop. Frankfurt am Main, Germany, December 1996.

[15] Wireless Application Protocol. Conformance Statement, Compliance Profile and Release List. WAP forum, April 1998. URL: `http://www.wapforum.com`.

[16] Wireless Application Protocol. Architecture Specification, WAP forum, April, 1998. URL: `http://www.wapforum.com`

[17] Wireless Session Protocol Specification. WAP forum, URL: `http://www.wapforum.com`, 1999

[18] Jzef Wozniak. Analysis of a New Hybrid PRMA-Type Channel Access Scheme with Frequency Hopping for Cellular Mobile Systems. IFIP TC 6 - Task Group Wireless Communications 2nd PWC Workshop. Frankfurt am Main, Germany, December 1996.

A Comparative Study on Distributed Location Management Strategies in Wireless Networks

Hoang Nguyen-Minh, Harmen R. van As
Vienna University of Technology, Institute of Communication Networks
Favoritenstrasse 9/388, A-1040, Vienna, Austria
Tel.: + 43-1-58801-38815, Fax: + 43-1-58801-38898
Email: Hoang.Nguyen-Minh@tuwien.ac.at

Keywords: TINA, UMTS, Distributed Location Management.

Abstract: Location management is an essential process in future mobile communication networks. An important issue is an efficient management of the location database. In this paper, the next generation mobile communication networks are proposed to integrate with a TINA-compliant architecture enabling to handle that kind of mobile-specific processes. We consider a distributed location database architecture for location management performing the following strategies: HLD (home location database only), HLD-VLD (HLD with a visited location database), or VLD-CLD (HLD and VLD with a cache location database). This paper discusses design, modelling and the comparison of the mentioned distributed location management strategies. The performance measures used for comparison are communication cost (signalling messages), computational cost (database accesses) and average total cost. For the performance analysis, we assume that the cost of updating a cache pointer and a user profile at CLD/VLD is equal. Results show that the combination of replication with caching scheme (VLD-CLD) performs better than the replication scheme (HLD-VLD) for a very wide range of call-to-mobility ratio (CMR).

1. INTRODUCTION

Location management is an essential process in mobile communication networks supporting location independent communications to users [1]. Location management is concerned with the issues of tracking and finding the mobile terminals (MT) in order to allow roaming within the network coverage area. The third generation mobile communication network, Universal Mobile Telecommunication System (UMTS), has adopted the functional type architecture of Intelligent Networks (IN) and distributed database architecture has been proposed [2]. It is still an open question whether the designated concepts of IN will be sufficient to provide the required powerful mechanisms needed for service provisioning, connection management, as well as mobility management in multi-service provider domains. TINA (Telecommunications Information Networking Architecture) proposes a framework for future telecommunication systems based on distributed and object-oriented computing that enables easy implementation of a distributed database in a distributed processing environment for location management.

Three strategies for location management in wireless networks are proposed in this contribution: HLD (home location database only), HLD-VLD (HLD with a visited location database), and VLD-CLD (HLD and VLD with a cache location database). We give the protocols of location updating/registration and call delivery procedures for these distributed databases. We compare those algorithms and show the effectiveness of the algorithm of VLD-CLD and HLD-VLD schemes by an analytical method.

The rest of the paper is organized as follows. In Section 2, we describe the system model and the distributed location database architecture. In Section 3, we give some of the strategies for location update and call delivery. We present an analysis of these algorithms and compare their performance in Section 4. Finally, some concluding remarks are given in Section 5.

2. SYSTEM MODEL

The basic system model for wireless communication networks using a TINA-compliant model, similar to the one presented earlier [5], is shown in Figure 1. Essentially, each cell has a base transceiver station (BTS) to which the mobile terminals of the cell communicate through a wireless link. Each BTS is connected to a mobile switching center (MSC) through a wired network. The BTSs and MSCs are modelled as sub-networks and layer networks that are under the control of the layer network domains (LND) and the connectivity provider domain. The cells are aggregated into contiguous geographical regions called Location Areas (LA). An LA is the granularity at which the network keeps track of the locations of the MTs. An object located in the MT called Terminal Location Management (TLM) performs location updating [5]. Each MSC has an object called Location Manager (LM), which performs the location update/registration and search procedures. The LM has access to the home location database (HLD) within the LND, which is used to store location and service information for each registered user of the wireless network. Each MSC has a cache location database (CLD) or a Visited location database (VLD) performing the schemes of caching or replication. The CLD/VLD also plays a role in handling call control information, authentication and billing. Note that a CLD/VLD has a geographical dependency and is coupled to an MSC and a fixed set of BTSs (grouped into one or several LAs). The CLD and VLD architectures can also be generalized into a distributed database design. In this distributed design, user profiles are partitioned and stored in multiple databases at different physical locations within the MSC. A TINA-compliant wireless network can support a distributed processing environment (DPE), a middleware kernel like CORBA's ORB, on which the computational objects are running and interacting.

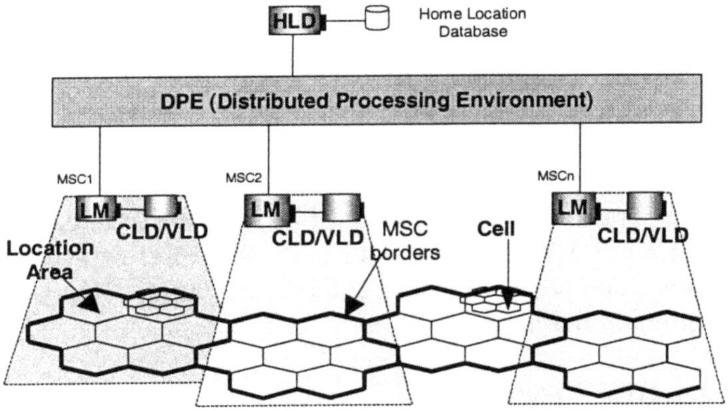

Figure 1. Wireless Network Architecture based on TINA-compliant model

3. LOCATION MANAGEMENT PROTOCOLS

3.1 HLD-only

In this strategy, *only* the home location database (HLD) keeps the location information of its enrolled MTs. Thus, one database query is enough to deliver the call. Whenever an MT moves to another LA, a location update request is delivered only to its home HLD to update the current LA address. For the call delivery procedure, the location information of a called MT is retrieved from the called home database to determine the current LA address and a paging process is started. This strategy reduces the number of database accesses and simplifies the call delivery procedure [3]. The location update procedure and call delivery procedure are given in Figures 2a and 2b, respectively.

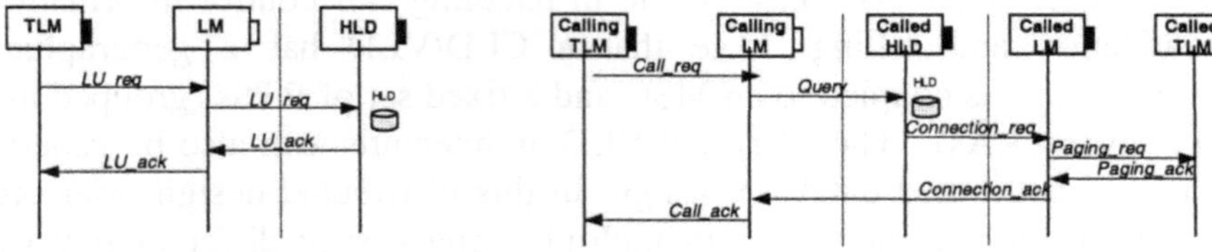

(a) Location update (b) Call delivery

Figure 2. HLD-only location management

3.2 HLD-VLD

In this section, we introduce a *simple replication* (SR) scheme called HLD-VLD. It is based on distributed visited location databases (VLDs) with a replication function that enables copying the user information from the HLD to the VLD in the visited MSC corresponding to the user's movement. The VLD temporarily stores the user profiles, and the location information (the current LA) of MTs roaming in its area. The HLD permanently stores the user profiles and points to the VLD that is in charge of the LA in which the MT is currently roaming.

3.2.1 Location Update (LU)

When an MT leaves an LA and enters another LA, it initiates a location update procedure. There are two types of location update procedure: intra-MSC and inter-MSC.

- **Intra-MSC LU**

When an MT enters a new LA in the same MSC, the updating procedure will be done locally at the current VLD of the MSC, as shown in Figure 3a. The cost due to the location update can be reduced because the MT does not need to register at its home HLD.

- **Inter-MSC LU**

This procedure will be triggered whenever the MT enters a new LA in the new MSC as shown in Figure 3b. The LM determines that it is an inter-MSC procedure by the identifier of the MSC, and then forwards the request to the home location database at the HLD. The new location information of the MT is updated in the HLD, and if authentication successful is, an acknowledgement of the update request is returned to the current VLD. The old VLD is also informed to remove the record of the MT and return the acknowledgement message to the HLD.

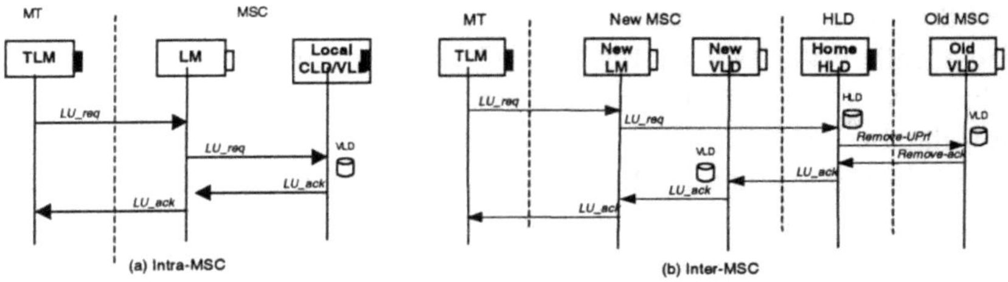

Figure 3. HLD-VLD location update procedure

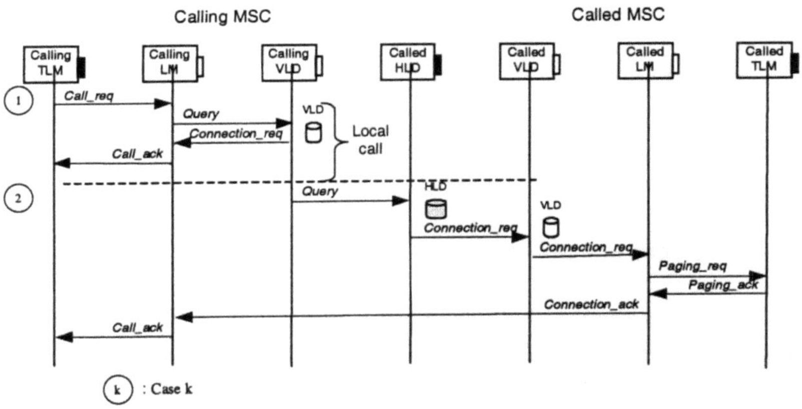

Figure 4. HLD-VLD call delivery procedure

3.2.2 Call delivery procedure

For the call delivery protocol, the location information of a called MT is first retrieved from the VLD of the calling MT to check whether the called and calling parties are in the same MSC. If so, it is a *local call* delivery and the call will be routed without any HLD access (*case 1* in Figure 4). If not, the HLD is queried for the location information of the called MT. The HLD contains a pointer to the called VLD in whose associated LA the called MT is currently located, and launches the query to that VLD. The VLD in turn queries the called LM to determine whether the user terminal is capable of receiving the call (by performing a paging procedure). The called LM returns a routable address back to the calling LM. At this point the connectivity provider domain can route the call by the connection procedure which is in detail described in [5].

3.3 VLD-CLD

In this section, we introduce a *simple replication with caching* (SRC) scheme called VLD-CLD. Each MSC is equipped with a visited location database (VLD), as mentioned in the previous section and a local cache location database (CLD) that maintains temporally location caching of called MTs originated from their MSC. Location caching for a given user should be employed if a large number of calls originate for that user from that MSC, related to the user's mobility. The cache information is kept at the CLD/MSC from which the calls originate. Location caching involves the storages of location pointers at the originating MSC that point to the VLD called the *pointed VLD* (and the associated MSC) where the MT is currently registered.

3.3.1 Location update

The location update procedure for this scheme is similar to the HLD-VLD scheme in Section 3.2.1.

3.3.2 Call delivery procedure

Each MSC equipped with a cache location database (CLD) that maintains temporarily pointers to pointed VLDs and a visited location database (VLD) that stores the user profile replicas of users visited in its LA. Note that the cache information is kept at the calling CLD/MSC from which calls originate. On receipt of a call request (*call_req*) message, the calling LM checks its local VLD whether the information of the called party is available locally.

If the calling party and the called party are in the same MSC, and the called user profile is stored at the VLD, we have a *local call procedure* as shown in *case 1* of Figure 5. In the local call, the called user location is received locally at the VLD; hence it completes a call fast and needs less signaling traffic and fewer database accesses. This is useful due to the fact that many call patterns in personal communication services today have the locality between the calling and called parties.

A *remote call delivery* situation occurs when the calling and called parties are in different MSCs. The calling LM then checks the local CLD for location information of the called MT. If a cache entry exists and the pointed VLD is queried, two situations are possible. If the called MT is still registered at the LA of the pointed VLD, we have a hit case in Figure 5 (*case a*).

The called MT is paged and the routing address is returned to the calling LM. Otherwise, the pointed VLD returns a *cache miss* in which the local CLD queries the HLD of the called party and then caches the location information of the called MT (the pointer to its pointed VLD) as shown in *case b* of Figure 5.

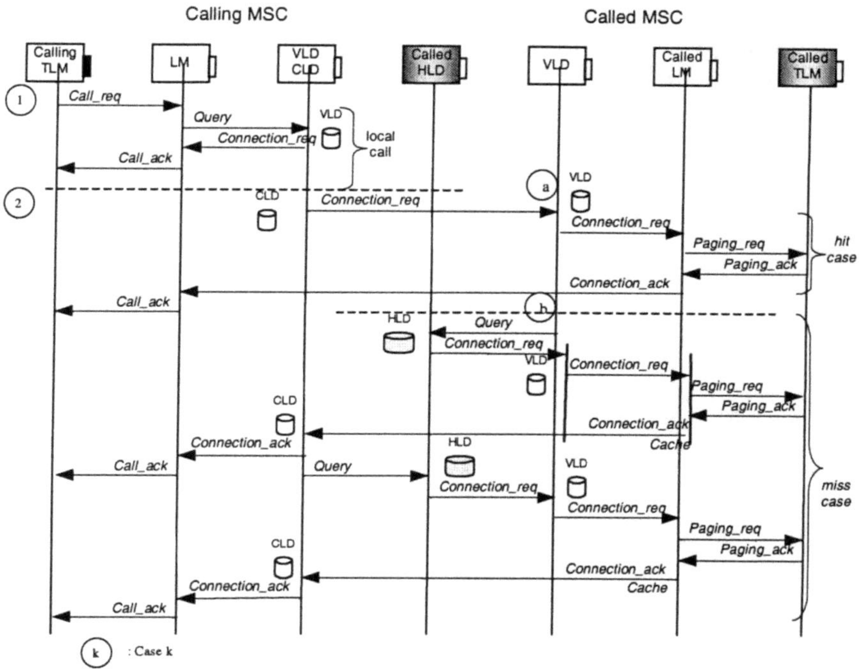

Figure 5. SRC (VLD-CLD) call delivery procedure

4. PERFORMANCE ANALYSIS

4.1 System Analysis

We assume that the density of users is uniform throughout the area, that the direction of motion with respect to the border is uniform on [0, 2π), and that all the cells are of the same shape and size and together form a contiguous area. Let t_c and t_m be independent and identically distributed random variables representing the call arrival time and the LA residence time. We assume t_c and t_m to be exponentially distributed with the rate of λ_c and λ_m, respectively. According to [7], the border-crossing rate out of a circular LA and MSC areas, denoted by ν and μ, with v is MT's average speed, are given by

$$\nu = \frac{\pi v}{4r_{LA}}, \text{ and } \mu = \frac{\pi v}{4r_{LA}} \tag{1}$$

where r_{LA} and r_{MSC} are the radius of the circular area of the LA and the MSC accordingly. Note that an MT that crosses an MSC will also cross an LA border. So, the border-crossing rate for which the MT still stays in the same MSC

$$\lambda = \nu - \mu = \frac{\pi v}{4}\left(\frac{1}{r_{LA}} - \frac{1}{r_{MSC}}\right) \tag{2}$$

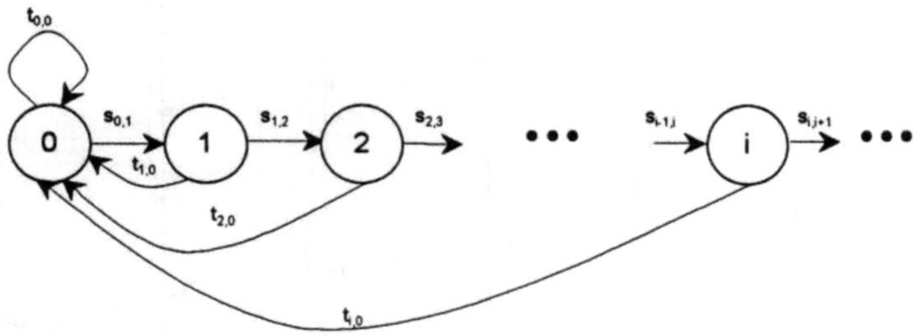

Figure 6. Imbedded Markov chain model

Figure 6 shows an imbedded Markov chain model, which describes the location update process of an MT. The state of the imbedded Markov chain, i, is defined as the number of LAs in the same MSC that the MT has passed by. The state transition rate $s_{i,i+1} = \lambda$ represents the MT moving rate from state i to state $(i+1)$ to the neighbouring LA in the same MSC. The transition rate that the MT moving (from state i to state 0) to another LA out of the MSC, denoted by $t_{i,0} (i \geq 0)$, is μ.

We assume p_i to be the equilibrium state probability of state i. The expression for p_i $(i \geq 0)$ in term of p_0 is

$$p_i = [\lambda/(\lambda+\mu)]^i p_0 = (1-p_0)^i p_0 \tag{3}$$

Using the law of total probability, the equilibrium state probability of state 0 is obtained as

$$p_0 = \frac{\mu}{\lambda+\mu} = 1 - \frac{\lambda}{\lambda+\mu} = 1 - \rho = \frac{r_{LA}}{r_{MSC}}, \text{ with } \rho = \frac{\lambda}{\lambda+\mu} \tag{4}$$

4.2 Cost Analysis

The location tracking cost is divided into two components: (1) *the update cost*, U_x, which incurred in completing the location update procedure, (2) the *searching cost*, S_x, which incurred in completing the call delivery procedure ($x \in \{HLD\text{-}only, SR, SRC\}$). We consider communication (routing, sending, receiving messages) and database processing as our basic measures of cost. The protocols involve the setup of dedicated channel and the call delivery procedure includes a paging procedure for all schemes, thus we ignore these costs for all calculations. Some costs and parameters used to analyse the analytic model are summarized in Table 1.

Table 1. List of cost parameters

Parameters	Meaning	Value
D_h	The cost for a query or an update of the HLD	αD_{cv}
D_{cv}	The cost for a query or an update of the CLD/VLD	γN_{lo}
N_{re}	The cost for a signalling message through the remote link to another MSC	βN_{lo}
N_{lo}	The cost for a signalling message through the local link within the MSC	1
C_{intra}	The cost of an intra-MSC location update procedure	$2N_{lo}+D_{cv}$
C_{inter}	The cost of an inter-MSC location update procedure	$N_{lo}+4N_{re}+D_{cv}+D_h$

4.2.1 Update Cost

According to the location update protocol for the intra-MSC and inter-MSC cases in Figure 3, the expression of the average movement costs for SR and SRC schemes are given by

$$U_{SR} = U_{SRC} = p_o C_{inter} + C_{intra} \sum_{k=1}^{\infty} k p_k = (1-\rho) C_{inter} + \frac{\rho}{1-\rho} C_{intra} \tag{5}$$

The update cost for the HLD-only scheme is simply received by

$$U_{HLD\text{-}only} = (2N_{re} + D_h)\left[(1-\rho) + \frac{\rho}{1-\rho}\right] \tag{6}$$

4.2.2 Searching Cost

The searching cost for HLD-VLD and VLD-CLD depends upon the probability of a calling and called parties are in the same MSC, q, the cache-hit ratio, δ, as well as the cost of querying the pointed VLD that store in the cache CLD. The cache-hit ratio is defined as the relative frequency with which the cached pointer correctly points to the user's location when it is consulted [6]. Table 2 summarises all possible events, their corresponding probabilities, and costs for the searching procedures of all the schemes.

Table 2. Searching activities, probabilities and costs

Scheme	Description		Probability	Symbol	Cost
HLD-only	--		1	$S_{HLD\text{-}only}$	$3N_{re}+D_h$
SR (HLD-VLD)	local call		$p_{SR,1}=q$	$S_{1,SR}$	$2N_{lo}+D_{cv}$
	remote call		$p_{SR,2}=(1-q)$	$S_{2,SR}$	$2N_{lo}+3N_{re}+2D_{cv}+D_h$
SRC (VLD-CLD)	local call		$p_{SRC,1}=q$	$S_{1,SRC}$	$2N_{lo}+D_{cv}$
	remote call	hit	$p_{SRC,2}=\delta(1-q)$	$S_{2,SRC}$	$2N_{lo}+2N_{re}+2D_{cv}$
		miss	$p_{SRC,3}=(1-\delta)(1-q)$	$S_{3,SRC}$	$3N_{lo}+3N_{re}+3D_{cv}+D_h$

The average cost per unit time can be expressed as the product of the cost and the rate of their occurrence as

$$S = \lambda_c \sum_i p_i S_i = \lambda_c S_x \tag{7}$$

The total cost per unit time is the combined cost of location update and searching, denoted as T, which is the sum of the two costs, given as

$$T = U + S = \lambda_m U_x + \lambda_c S_x \tag{8}$$

4.3 Comparative Results

In evaluating and comparing the cost of these strategies, we consider the call to mobility ratio (*CMR*) as the ratio of the call arrival rate to the mobility rate and the cache-hit ratio as

$$CMR = \frac{\lambda_c}{\lambda_m} = \frac{\delta}{1-\delta} \tag{9}$$

In order to be able to estimate the total cost per unit time, the call arrival rate λ_c, and the rate at which the MT moves between LA, λ_m, are needed. We use the call to mobility ratio (*CMR*) to study the performance of these schemes and compare the cost of the strategies by the ratio of the total costs (*TCR*) described as follows

$$TCR = \frac{T}{T_{ref}} = \frac{U_x + CMR * S_x}{U_{ref} + CMR * S_{ref}} \tag{10}$$

For the analytical results given in this section, it is assumed that the cell is about 150-200m of radius, whereby LA and MSC cover around 100 cells and 100 LAs, respectively. From equations (5) and (6), we get cost reduction of 10 times for the location update procedure referred to the HLD-only scheme. The methodology of evaluation used is to establish a common unit of measure for all cost terms, for example time delay. We use the sets of values of coefficient parameters α, β, and $\gamma \in (1,2,3)$ as shown in Table 3. The value of N_{lo} is normalized to one since it can be seen as the lowest among the other costs. Parameter sets 6 and 8 capture the cases when it is significantly more expensive accessing the D_h than D_{cv}. Figure 7 plots the variation of the normalized total cost with a wide range of *CMR* (0.01 to 100). We used the value of q=0.25 so that CMR gets a threshold for caching of being beneficial [6]. It can be seen that (1) the cost improvement of the VLD-CLD/HLD-VLD schemes have meaning only if the access cost to HLD is greater than that to VLD/CLD; (2) the VLD-CLD/HLD-VLD schemes always result in higher performance than the HLD-only strategy; (3) the VLD-CLD performs better than the HLD-VLD for a very wide range of call-to-mobility ratio, especially when CMR is greater than one.

Table 3. Coefficient parameters

Class	N_{lo}	N_{re}	D_{cv}	D_h		Class	N_{lo}	N_{re}	D_{cv}	D_h
1	1	1	1	1		5	1	2	2	4
2	1	2	1	1		6	1	2	2	6
3	1	2	1	2		7	1	3	3	6
4	1	2	2	2		8	1	3	3	9

5. CONCLUSION

In this paper, the next generation mobile communication network is proposed to integrate with a TINA-compliant architecture enabling to handle that kind of mobile-specific processes. This paper addresses design, modelling and comparison of a distributed location management. In the proposed network model, three mechanisms have been applied: distributed home location database, caching and replication strategy for handling the location management procedures. We use the local cache (CLD) and replication (VLD) location databases to reduce network signalling traffic and database updating as well as querying delay during location registration and call delivery procedure. The numerical results show that both strategies, VLD-CLD and HLD-VLD, can significantly reduce the total cost compared to the HLD-only scheme. Results also show that the replication with caching strategy performs better than the replication strategy. This paper also presents a simple methodology for evaluating some basic location management algorithms for distributed location databases based on a TINA-compliant architecture for wireless communication networks.

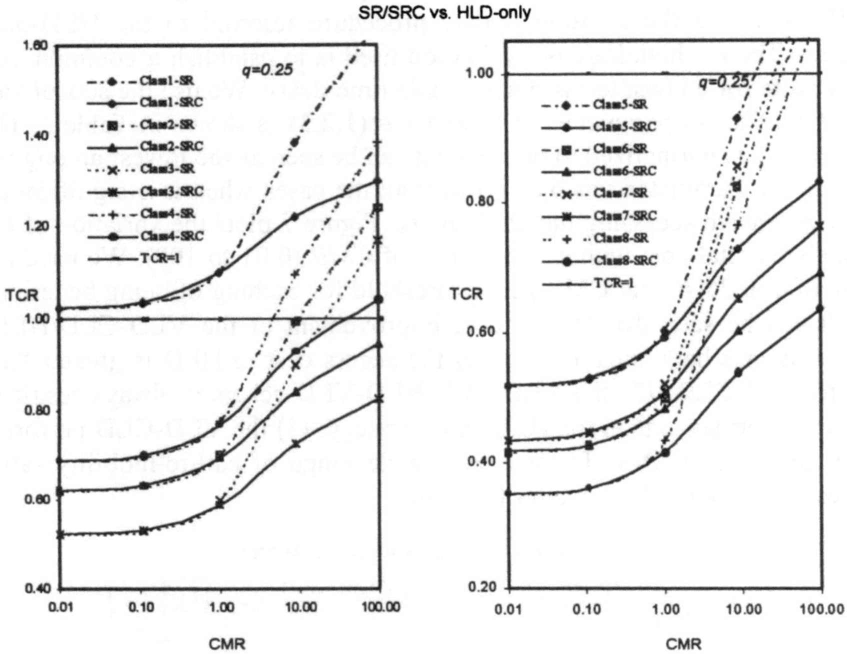

Figure 7. Total cost Ratio (TCR)

REFERENCES

[1] I.F. Akyildiz, J. McNair, J.S.M. Ho, H. Uzunalioglu, and W. Wang, "*Mobility Management for Next Generation Wireless Systems*", Proc. IEEE, Vol. 87, No. 8, Aug. 1999, pp. 1347-1384.

[2] C. Eynard, M. Lenti, A. Lombardo, O. Marengo, and S. Palazzo, "*A methodology for the performance evaluation of data query strategies in UMTS*", IEEE JSAC, Vol. 13, No. 5, June 1995, pp. 893-907.

[3] B.C. Kim, J.S. Choi, C.K. Un, "*A new distributed location management algorithm for broadband Personal Communication Networks*", IEEE Trans. on Vehicular Technology, Vol. 44, No. 3, Aug. 1995, pp. 516-524.

[4] K. Wang, J. Huey, "*A cost effective distributed location management strategy for wireless networks*", Wireless Networks, Vol. 5, No. 4, Aug. 1999, pp. 287-297.

[5] N.M. Hoang, H. R. van As, "*Connection and Location Management based on a TINA-compliant Architecture for UMTS*", Smartnet'99, Thailand, Nov. 1999.

[6] R. Jain, Y.-B. Lin, C. Lo, S. Mohan, "*A Caching strategy to reduce Network impact of PCS*", IEEE JSAC, Vol. 12, No. 8, Oct. 1994, pp. 1434-1444.

[7] J. Schuringa, "*Performance Modeling of Location Management Strategies in Mobile Networks*", Master Thesis 1995, Department of CS, Univ. of Twente.

Resource Allocation in Cellular Wireless Systems

Villy B. Iversen and Arne J. Glenstrup
Department of Telecommunication, Building 371 Technical University of Denmark, DK–2800 Kongens Lyngby,Tel: (+45) 4525 3648, Fax: (+45) 4593 0355
e–mail {vbi,panic}@tele.dtu.dk

Key words: Mobility, modelling, microcell, macrocell, UMTS

Abstract: In mobile communications an efficient utilisation of the channels is of great importance. In this paper we describe the basic principles for obtaining the maximum utilisation and study strategies for obtaining these limits. In general a high degree of sharing is efficient, but requires service protection mechanisms to guarantee the Quality of Service for all services. We study cellular systems with hierarchical cells, and the effect of overlapping cells, and we show that by call packing we obtain a very high utilisation. The models are generalisations of the Erlang-B formula, and include general arrival processes, and multi-rate (multi-media) traffic for third generation systems.

1. INTRODUCTION

The modelling of a cellular communication system is usually divided into modelling of (a) traffic, (b) structure, and (c) strategy. In comparison with plain old telephone systems the problems become more complex as e.g. the number of channels available depends on the position of the subscribers, the subscribers are moving, and the intelligence available for decision-making is very high.

2. CELL DIMENSIONING

We can model a basic cellular network of N cells analytically by considering a state space in which each state is indexed by a vector $[i_1,\ldots,i_N]$ where i_k is the number of customers being served by cell k. New calls, call terminations and call handovers are modelled by Poisson processes with state-dependent intensities, causing single state transitions. This model is generally non-trivial to solve, due to the handover transitions, but it has been shown that it can be approximated by a model in which the effect of handovers have been transformed into an increase in the birth and death rates [1, 3]. This latter model, which has product form, can be solved and subsequently used to compute various performance measures (blocking probabilities for new calls and handovers, utilisation etc.). The approximation is not exact, but can be considered as a worst-case scenario when dimensioning the cells of a mobile network.

Measurements (e.g. [10]) have confirmed that new calls arrive according to a Poisson process with slow variations during the day. It has also been shown that handover traffic is more smooth than the Poisson process [9].

All the models considered in the following have product form and are insensitive to the holding time distribution. Because of the product form we may apply the convolution algorithm for loss systems, first published in [4]. In a network with direct routing we have product form between the routes. The convolution algorithm allows for class limitation by truncating the state space and for minimum allocation by aggregation of states [5]. The models are valid for any state-dependent Poisson process and multi-rate traffic. More details are given in [6].

3. OVERLAY & UNDERLAY STRUCTURE

The performance of cellular mobile communication systems can be improved significantly by introducing macro cells. If a call (either a new or a hand over call) attempts to establish a connection in a micro cell but all the channels of that cell are busy, it may try to establish the connection in an overlaid macro cell. Figure 1 shows a two-level system with N microcells and one macrocell. Micro-cell number i has n_i channels of its own. The macrocell has a total of m channels, but microcell number i may at most borrow m_i channels in the macrocell at a given point of time (class limitation). Thus the subscribers in a given microcell has a minimum allocation of n_i channels and a maximum allocation of $n_i + m_i$. This allows us to guarantee a certain grade of service. The macro cell is common to a number of micro cells, i.e. it acts as a shared resource for a number of micro cells. The multi-layer structure is equivalent to that of a classical overflow system. Two call management strategies for operating macro cells exist [8]:

- *Without rearrangement*, i.e. once a call has established a connection in a channel in a macro cell, it continues to utilize the channel until the call is terminated.

- *With rearrangement*, if a call is using a channel in a macro cell it will rearrange to the micro cell (where the call is located) when a channel in that particular micro cell becomes idle.

The rearrangement strategy increases the number of hand over operations and requires that the system keeps information about which micro cell a call belongs to.

The performance of multi-layer systems are calculated in the following way. Systems without rearrangement are similar to classical overflow systems and can be evaluated by well-known methods. If the rearrangement strategy is applied, the state transition diagram is reversible and blocking probabilities, utilisation, etc. are obtained using the reversibility, for instance by using the *convolution algorithm* [4].

3.1. EXAMPLE

We consider a network with 40 micro cells and one macro cell covering all the micro cells. Each micro cell has 40 channels. The number of channels in the macro cell is a variable. The termination rate of a call is one (the time unit is chosen as the average holding time) and the arrival rate of new calls is per time unit. A call in a micro cell tends to move to one of the neighbour cells with a constant rate of one. When changing cell the new cell is chosen among the neighbouring cells with equal probability. In order to reduce the blocking probability a number of channels is allocated to the macro cell.

Figure 2 features a plot of the total carried traffic versus the number of channels in the macro cell. If the call management strategy without rearrangement is applied the utilisation of the channels in the macro cell is close to one. If the rearrangement strategy is applied, the extra carried traffic per additional channel in the macro cell is *higher* than one erlang. This is due to the fact that when a call is blocked at the micro cell and gets a channel in the macro cell it will only remain in the macro cell until a channel in the micro cell is released. On average this time is only 1/40 of a holding time (if only one call is waiting for a channel in the micro cell). In this way the shared resource (the macro channel) is made available for new calls as soon as possible. Obtaining the same amount of carried traffic (or correspondingly the same blocking probabilities) without rearrangement requires a significantly higher number of channels in the macro cell. However, the increased utilisation of the micro cells implies that a higher number of calls are blocked in the micro cells but accepted in the macro cell. Thus the total number of rearrangements increases. If the number of channels in the macro cell becomes large the system starts to make unnecessary rearrangements. A rearrangement of a call from the macro cell to a micro cell k is unnecessary if no other micro cell requests to use the released channel in the macro cell. When the number of channels in the macro cell is large the utilisation is small and many unnecessary rearrangement occur.

4. OVERLAPPING CELL BOUNDARIES

Above, we considered one macrocell overlapping all microcells. We may also have overlap between the microcells. In Figure 3 subscribers in area 1 have access to n_1 channels, subscribers in area 2 access to n_2 channels, and subscribers in area 12 have access to $n_1 + n_2$ channels. Therefore, subscribers in area 12 will experience a smaller blocking probability than subscribers in the other areas. In an intelligent system we can freely hand-over calls in area 12 between the two base stations. macrocell to the microcell when a channel

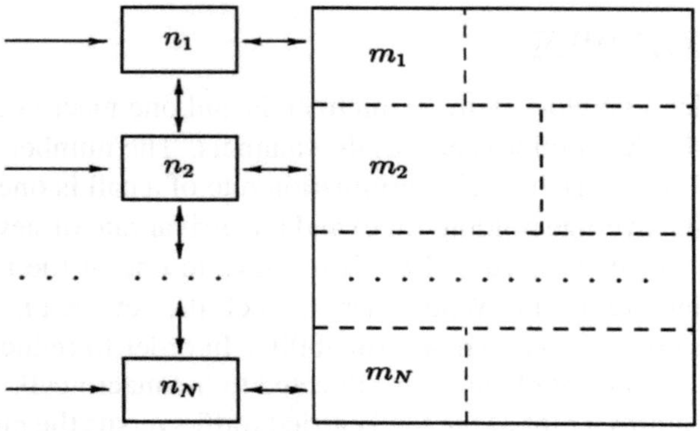

Figure 1 Model of a cellular system with N *microcells and one macrocell corresponding to a link model offered more traffic streams with minimum and maximum The model has product form.*

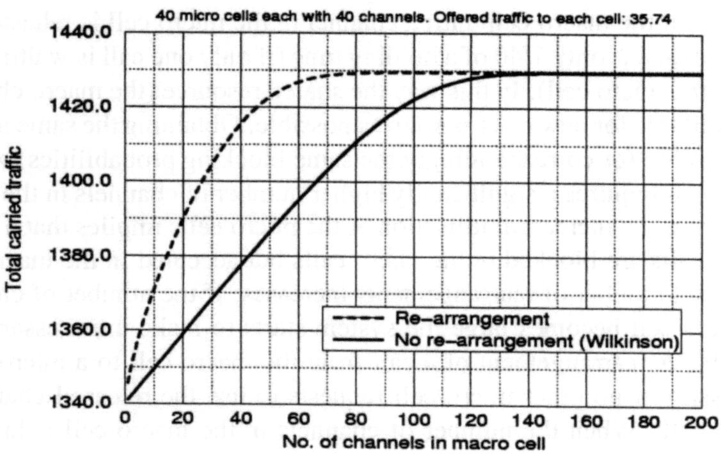

Figure 2 The total carried traffic as a function of the number of channels in the macro cell. By adding 20 channels in the macro cell we notice that the total carried traffic increases by 50 erlang.

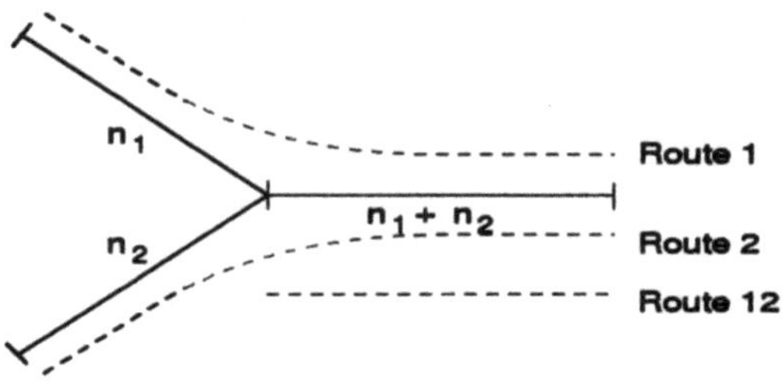

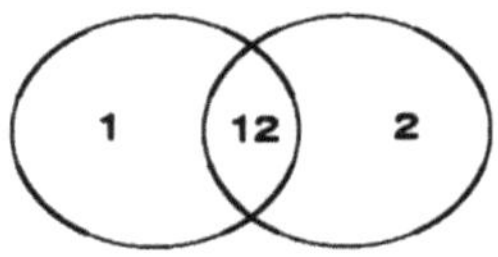

		Route 1	Route 2	Route 12
Link	1	1	0	0
	2	0	1	0
	12	1	1	1

Figure 3 Example of two overlapping cells and the equivalent circuit-switched network with direct routing.

becomes available. Denoting the number of existing connections in the area i by x_i we notice that we have the following restrictions:

$$0 \quad x_1 \quad n_1 \tag{1}$$

$$0 \quad x_2 \quad n_2 \tag{2}$$

$$0 \quad x_1 + x_2 + x_{12} \quad n_1 + n_2 \tag{3}$$

This is equivalent to the circuit switched communication network with direct routing shown in the Figure 3. We describe a network with direct routing by the link, the routes and the number of channels c_{ij} (element of the matrix) a route requires on each link. The models are valid for multi-slot systems with individual slot size on each link.

We notice that if all channels are busy in e.g. area 1, then we may hand-over a connection in the area 12 from base station one to base station two. Thus we assume optimal rearrangement (call packing). In a multi-cell systems this rearrangement may be necessary at several levels. Thus we assume that the system has global optimal intelligence. In Figure 4 we consider a system with three cells, which are mutually overlapping, so that subscribers in some areas may have access to two base stations, but not three. Therefore, subscribers in overlapping areas will experience a smaller blocking probability than subscribers in the separate areas. In an intelligent system with optimal packing we can rearrange calls in the overlapping areas from one base station to another. Thus we assume that the system has global optimal intelligence. The model with restrictions on the number of simultaneous calls is equivalent to a circuit switched communication network with direct routing [5] as shown in the Table

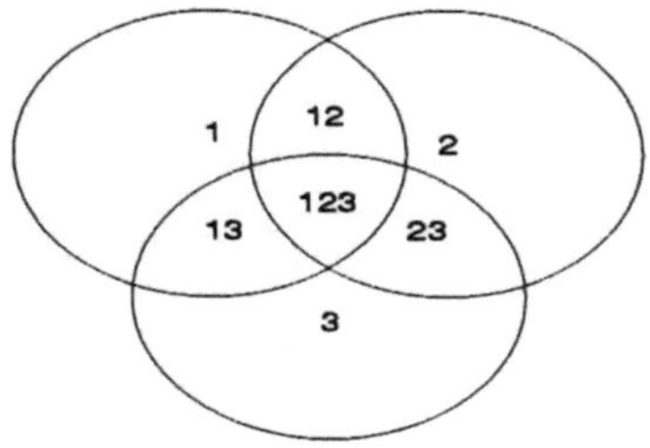

Route

Link	1	2	3	12	13	23	123
1	1	0	0	0	0	0	0
2	0	1	0	0	0	0	0
3	0	0	1	0	0	0	0
12	1	1	0	1	0	0	0
13	1	0	1	0	1	0	0
23	0	1	1	0	0	1	0
123	1	1	1	1	1	1	1

Figure 4 Example of three overlapping cells and the corresponding equivalent circuit-switched network with direct routing.

of Figure 4. A link corresponds to a restriction The number of routes becomes equal to the number of separate areas, whereas the number of links becomes equal to the number of restrictions, which is equal to the number connected areas (paths) we can built up from the distinguishable areas. If all N cells are overlapping the number of links become equal to 2^N 1, as we exclude the empty set. For the case considered in Figure 4 the number of routes becomes 6 and the number of links becomes 7 (in fact, one of the restrictions is superfluous).

For the case considered the carried traffic as a function of the overlapping is shown in Figure 5. We notice, that we have the same capacity as for full availability when the overlapping is greater than 20 %. This will be the case in most real systems. The model with overlay cell is included in this model has the overlay cell is a cell overlapping all other cells.

Evaluation methods: For small networks we have exact algorithms for evaluating the end to end blocking probability for each route, i.e. for each area. The convolution algorithm [4] allows calculation of both time congestion, call congestion, and traffic congestion for Multi-slot Binomial - Poisson - Pascal traffic.

As the number of routes and links for realistic systems (e.g. GSM) becomes very large, the exact methods are not applicable. Then numerical simulation and approximate methods as the Erlang reduced load (Erlang fixed point) methods has to be used. However for large networks, where a typical route

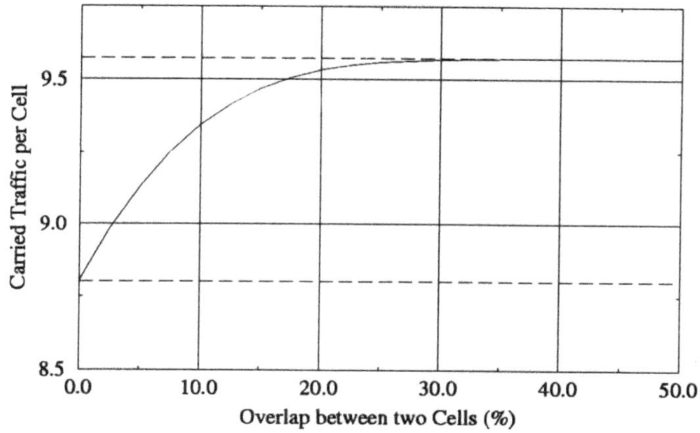

Figure 5 *System with three cells adjacent to each other.* A = 10 *erlang per cell,* n = 12 *channels per cell. 10 % overlap means that a cell has 10 % overlap with both of the two other cells, but there is no overlap between all three cells.*

may use 20 links, these methods are not very good.

Decentralized intelligence. In the above models we have assumed that the call packing is optimal. We may thus implement many successive rearrangements to move one idle channel from one cell to another. In e.g. DECT systems the intelligence is distributed to the individual handsets. Thus it is possible to let the handset, which knows the state of the channels at the local base station in the area, make local decisions based on the local information. Thus a handset may hand-over a call from a base station with all channels busy to a base station with idle channels. However, this strategy will not always be globally optimal (but we know the optimal reference value). A random model will choose an idle channel at random. These models can (only) be evaluated by simulation.

5. UMTS

Until this point, we have been considering 2nd generation mobile systems, i.e. the GSM and DECT systems. However, in the near future we will see the introduction of the 3rd generation mobile systems, UMTS (Universal Mobile Telecommunications System). The main extension in UMTS is the introduction of packet-switches connections for data transfer, and a design that is able to supply a greater bandwidth than GSM, typically 384 Kbps, and up to 2 Mbps. However, the increased bandwidth is not evenly distributed: the radio interface is designed so that the available capacity decreases with the distance of the mobile terminal from the base station. High-speed data transfer requires the largest bandwidth, and a voice connection the least, so to prevent a single terminal far from the base station using up all the capacity, connections there will be restricted to low bandwidths. Thus we can model a UMTS cell as a number of concentric rings with different available services, as shown in Figure 6 [2]. The packet switched services may be transformed to circuit switched traffic by using effective or equivalent bandwidths [7].

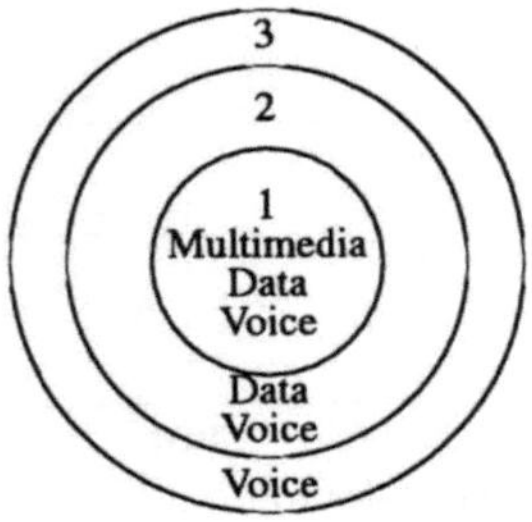

Service	*Bandwidth*
Multimedia	2 Mbps
Data	384 Kbps
Voice	16 Kbps

Figure 6 An example of a UMTS base station: in the outer rings only the low-bandwidth services are available

5.1. TRAFFIC MODELLING RESULTS

We model a UMTS cell carrying 20 channels in total, with only 10 and 5 being available in ring 2 and 3. Each call requires 1, 2 or 5 channels for a voice, data or multimedia connection, respectively. Table 1 shows the blocking probabilities for the services and the total carried traffic in the cell when more channels are available in the outer rings (10–13 in ring 2, and 5–8 in ring 3).

Normally, however, the number of channels available in the rings cannot change, and with the basic setup of 20/10/5 available channels, this causes voice call blocking in ring 3 to be 4 times as high as voice calls in ring 1. If we want to even out the blocking, we can enforce a reservation strategy whereby a number of "private" channels are reserved exclusively for calls in ring 2 and 3. Table 2 shows how blocking and carried traffic change under this strategy. Note that reserving one channel in each outer ring results in an *increase* in the total carried traffic; this is due to a carried traffic increase in ring 2 greater than the carried traffic decrease in ring 1. Figure 7 is a plot of how the blocking depends on the number of reserved channels in the outer rings.

6. CONCLUSIONS

By exploiting the capabilities of digital systems we are able to obtain a high utilisation of the radio channels. In particular systems with over-lapping cells and overlaid cells are able to manage local overload and at the same time guarantee a certain grade-of-service. If we are able to rearrange calls, then we are

	Ring 1			Ring 2		Ring 3	Carried
	M-media	Data	Voice	Data	Voice	Voice	traffic
Offered traffic	0.5	1	2	1	2	2	
Channels/call	5	2	1	2	1	1	
Max channels	20			10		5	11.350085
Blocking	0.176014	0.047955	0.020968	0.143045	0.060094	0.082878	
Max channels	20			11		6	11.476073
Blocking	0.185510	0.051442	0.022568	0.108263	0.045381	0.052423	
Max channels	20			12		7	11.548188
Blocking	0.191657	0.053890	0.023737	0.084987	0.035924	0.037797	
Max channels	20			13		8	11.586782
Blocking	0.195412	0.055411	0.024497	0.071045	0.030479	0.030911	

Table 1 Blocking probabilities and carried traffic under four different channel assignments

	Ring 1			Ring 2		Ring 3	Carried
	M-media	Data	Voice	Data	Voice	Voice	traffic
Offered traffic	0.5	1	2	1	2	2	
Channels/call	5	2	1	2	1	1	
Max channels	20c			10c		5c	11.350085
Blocking	0.176014	0.047955	0.020968	0.143045	0.060094	0.082878	
Max channels	18c			9c + 1p		4c + 1p	11.431453
Blocking	0.189841	0.052634	0.023090	0.108590	0.045122	0.068038	
Max channels	16c			8c + 2p		3c + 2p	11.409544
Blocking	0.216260	0.061221	0.026996	0.092638	0.038386	0.055663	
Max channels	14c			7c + 3p		2c + 3p	11.267149
Blocking	0.265696	0.077568	0.034512	0.089343	0.036827	0.046055	
Max channels	12c			6c + 4p		1c + 4p	10.985059
Blocking	0.349320	0.105625	0.047763	0.090701	0.036965	0.039766	

Table 2 Blocking probabilities and carried traffic under five different partitions of 20 channels into (c)ommon channels and (p)rivate channels reserved for a specific ring

able to evaluate the systems. For systems with fixed slot-assignment the exact solution is based on a very large number of linear equations. For multi-rate systems we have to protect wide-band traffic by using trunk reservation or class limitation.

References

[1] Christiansen, C. & Iversen, V.B. & Nasr, S. (1993): Product form solutions for cellular mobile communication systems. NTS 11, 11th Nordic Teletraffic Seminar, Stockholm 1993. 8 pp.

[2] Dekocker, S. (1999): Traffic problems in cellular wideband systems. Master's thesis. Dpt. of Telecommunication, Technical University of Denmark. 98 pp.

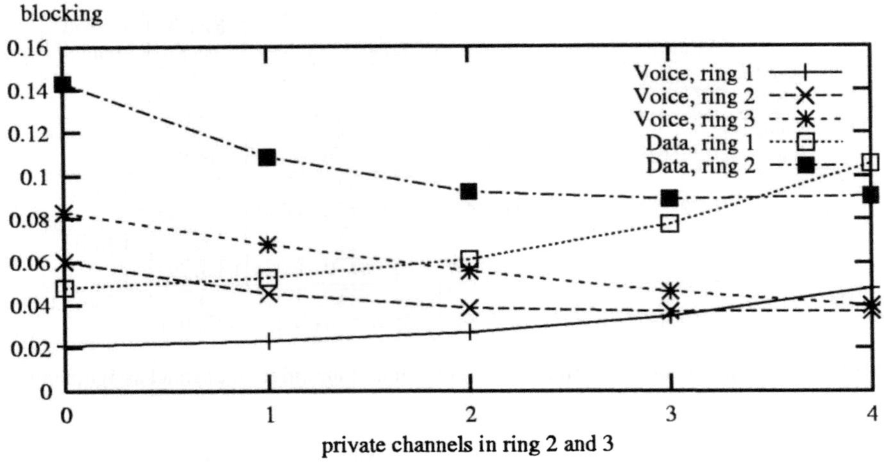

Figure 7 Blocking as a function of the number of private channels in the outer rings

[3] Everitt, D. (1991): *Product Form Solutions in Cellular Mobile Communication Systems*, Teletraffic and Datatraffic in a Period of Change, ITC–13, A. Jensen & V. B. Iversen (editors), North Holland, 1991, pp. 483–488.

[4] Iversen, V. B. (1987): *The Exact Evaluation of Multi–Service Loss Systems with Access Control.* Teleteknik, English ed., Vol. 31 (1987): 2, 56–61.

[5] Iversen, V. B. (1995): Traffic Engineering of Cellular Mobile Communication Systems. ITC Regional Seminar in Bangkok, November 28 – December 1, 1995. 10 pp.

[6] Iversen, V.B. (2000): *Teletraffic Engineering*. Chapter 11: Multi-dimensional loss systems. 61 pp. Dpt. of Telecommunications, Technical University of Denmark, 2000.

[7] Kelly, F. (1995): Notes on effective bandwidthsr, pp. 141–168 in Stochastic networks, theory and applications. Royal Statistical Society, London 1995.

[8] Lagrange, X. (1997): Multitier cell design. IEEE Communications Magazine, August 1997, pp 60–64.

[9] Rajaratnam, M. & Takawira, F. (1999): Non-classical traffic modelling and performance analysis of cellular mobile networks with and without channel reservation. University of Natal, Durban, South Africa. 39 pp. To appear in IEEE Trans. on Vehicular Technology.

[10] Smith, P.J. & Sathyendran, A. & Murch, A.R. (1999): Analysis of traffic distribution in cellular networks. 49th IEEE Vehicular Technology Conference, May 16–129, 1999, Houston, Texas, USA. Vol. 3, pp. 2075-2079.

Evaluation of Traffic Carried by ATM Wireless Access Link Controlled by MEDIAN Protocol

Andrzej Beben, Wojciech Burakowski and Piotr Pyda
Institute of Telecommunications, Warsaw University of Technology
Military Communication Institute, Zegrze
E-mail: abeben@tele.pw.edu.pl

Key words: wireless ATM, PRMA, CAC, QoS requirements, MEDIAN access protocol

Abstract: The paper points out on some limitations introduced by access points in wireless ATM network that have great impact on supporting *end-to-end* communication services with QoS requirements like CBR, VBR, ABR and GFR. Analysed access protocol is MEDIAN [5], which belongs to PRMA (Packet Reservation Medium Access) class and works under TDMA scheme. It appears that internal MEDIAN mechanisms (like frame structure, cell reservation mechanism and error protection) could radically reduce available link capacity designated for handling traffic with guaranteed QoS requirements. The formulas for evaluation this capacity are provided in the paper. Limitations for the CAC function are also discussed. Exemplary numerical results illustrating quality of transferred traffic are included.

1. INTRODUCTION

In 1996, the ATM Forum Wireless Working Group started work on the standards for the wireless ATM network (WATM) [12]. In order to adopt the ATM technology to a wireless environment, many problems should be solved because of high bit error rates in radio links and specific medium access protocols [1], [2]. However, solving these problems requires introduction of additional overheads necessary for access control, error protection and link organisation (frame structure, reservation mechanism) [3]. Anyway, the objective of wireless ATM network is to provide the same communication services as these supported by cable network (with similar QoS requirements) and this requires implementation of appropriate traffic control mechanisms also in the wireless access points [4].

The paper points out on some limitations introduced by access points in wireless ATM network having great impact on supporting *end_to_end* communication services with QoS requirements like CBR, VBR, ABR and GFR. Analysed access protocols, called PRMA (Packet Reservation Medium Access), belong to the family of TDMA class. More precisely, the MEDIAN protocol [5], as representative of this class is taken into account for our considerations. Effective traffic control (preventive and reactive) demands the precise knowledge of such parameters as available bandwidth and cell delay characteristics introduced by each hop (switch, access point) along the connection [9]. On the contrary to the cable link, the capacity available for the traffic on the wireless access point in much less than the link bit-rate while the cell delay and its variation is significantly larger. This is mainly due to additional overhead for access control, error protection and some limitations corresponding to the link organisation. Notice that available link capacity and cell delay characteristics in RAP (Radio Access Point) should be known a priori for performing CAC function. Evaluation of current available capacity is also strongly required for ABR service.

The paper provides an estimation of the available capacity in the RAP governed by the MEDIAN protocol. These considerations take into account the limitations corresponding to the internal access control protocol behaviour (frame structure, reservation mechanism, and error protection). Next, we point out on the conditions for the CAC performing. More specifically, estimation of the CDV (Cell Delay Variation) characteristics introduced by RAP is investigated. Recall that for some types on ATM connections, not only the CLR (Cell Loss Ratio) but also CDV value is also important. An excellent example of such connection is CBR connection, requiring rather small values of the CDV [8], [10]. On the other hand, correct estimation of the CDV values is necessary to avoid cell losses caused by the access protocol. Exemplary rules for tuning parameters assuming MEDIAN protocol are outlined. Finally, some exemplary numerical results are presented to illustrate consequences of admission traffics (of CBR and VBR types) up to full available capacity.

Organisation of the paper is as follows. Section 2 shortly describes the features of the MEDIAN protocol. The factors that result for reducing available bandwidth and related to the internal protocol mechanisms having great impact on traffic control are explained in section 3. Some numerical results showing the bandwidth limitations for the CAC function are shown in section 4. Finally, the conclusions are outlined in section 5.

2. BRIEF MEDIAN DESCRIPTION

In wireless ATM network a part of user terminals is attached to the fixed network by wireless links terminated in RAP unit, as depicted in Figure 1. For these terminals the access to the network is governed by special MAC protocol.

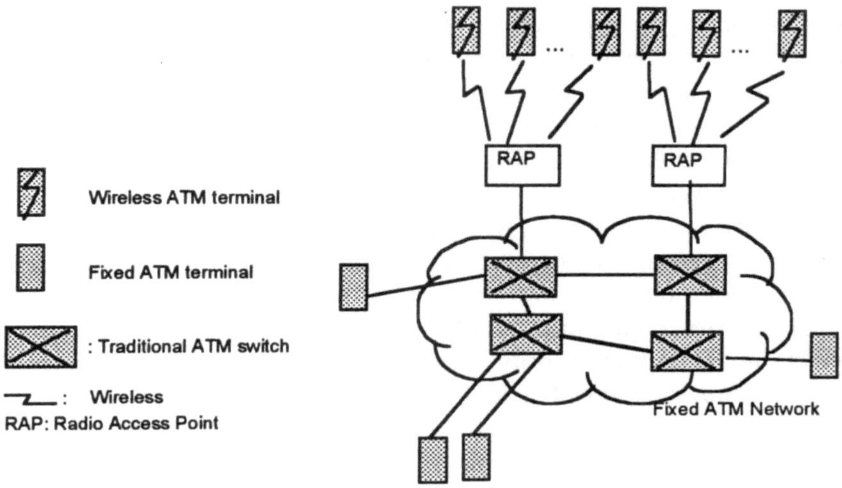

Figure 1. Wireless ATM network scenario

For the purpose of the wireless ATM network a number of new MAC protocols was recently submitted, which belong to the TDMA (Time Division Multiple Access) or CDMA (Code Division Multiple Access) class. In this paper we focus on the MEDIAN [5] protocol working under TDMA scheme. In fact, one can find also other protocols belonging to the considered class, like MASCARA [7] or SAMBA [6]. Detailed performance studies of these protocols are out of the scope of this paper. Anyway, all of mentioned protocols assume that the time slot allocation for a given connection is made dynamically. Therefore, some conclusions corresponding to the MEDIAN protocol behaviour can be the same for all protocols of the studied class.

In order to complete the paper, a short introduction to the MEDIAN protocol is provided below. Figure 2 depicts the frame format assumed for communication between mobile terminals and the base station. This frame consists of 64 slots, among them 61 are designated for the cell transferring while the rest are for sending control information (for synchronisation, broadcast and contention slots). Notice that this frame is for both up and down transmission. The number of slots designated for one of these directions depends on the current load conditions.

The base station dynamically allocates these 61 slots only for running connections. The updating process is made frame by frame on the basis of received reservation requests from end terminals for up transmission and from the base station for down transmission. In the case when the number of required reservations is smaller that 61, the rest slots in the frame are allocated according to a predefined pattern, which is assumed as equally distributed among connection being in progress. These slots are called polling slots. Notice that when the system is working under low load conditions then no reservations are generated since polling mechanism is sufficient to serve submitted traffic. On the contrary, high traffic load usually demands some reservations.

The active terminal sends its reservation requests only in these slots that is dedicated to it, using the piggybacking scheme. The mechanism governing the allocation of slots takes into account the number of received requests, their arriving times and priority of the connection. More precisely, in MEDIAN protocol pure priority scheme is not implemented. For each connection there is assigned a kind of priority that is expressed by the maximum allowed time (denoted as Δ_{max}) the cell can wait for sending to the base station. When this time terminates then the waiting cell is simply lost. Notice however that such mechanism does not guarantee that the cell from the higher priority connection could be served after a cell of lower priority. As a consequence, the applied EDD (Early Due to Date) scheduling with the above described cell discarding scheme does not support the possibility to reserve a fixed number of slots in consecutive transmission frames, what is extremely required e.g. for CBR connections. Therefore, cell transfer delay of consecutive cells depends on the traffic load conditions.

Similarly, the base station sends the requests for down transmission. In this case, the request for the slot reservation is generated when a cell from the network joins the base station. The requests arriving from the base stations and end terminals are served according to the expiration time values of particular cells. The cell with lower this value is served before. Anyway, inside one frame the cells for down transmission are served as the first.

More detailed description of the MEDIAN one can find in [5].

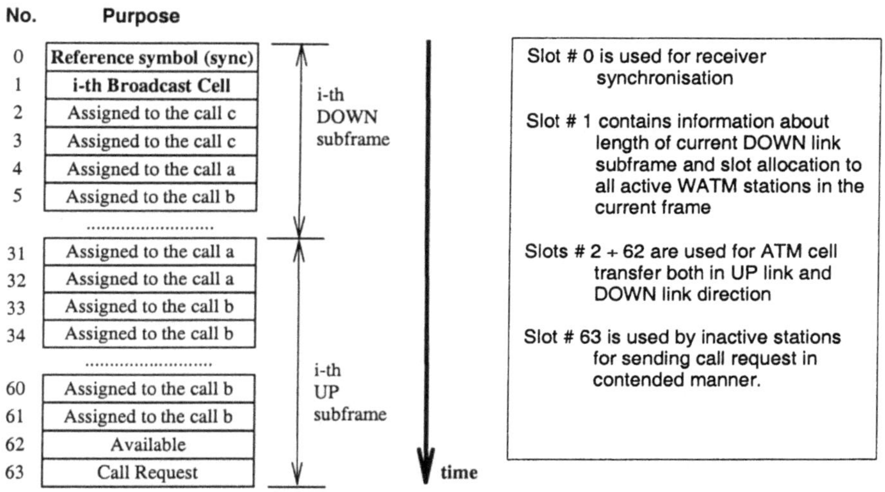

Figure 2 – Exemplary MEDIAN frame format in the case of three running connections, say a, b and c

3. GUARANTEED TRAFFIC SUPPORTED BY MEDIAN

In this section we investigate the limitations introduced by RAP controlled by MEDIAN protocol which have great impact on supporting *end_to_end* ATM communication services demanding bandwidth allocation, like CBR, VBR, ABR and GFR. Recall that effective traffic control (preventive and reactive) demands the precise knowledge of such parameters as available bandwidth and cell delay characteristics introduced by each hop (switch, access point) along the connection. On the contrary to the cable link, one can expect that capacity available for the guaranteed traffic on the wireless access point is much less than the link bit-rate while the cell delay and its variation is significantly larger. This is mainly due to additional overhead for access control, error protection and some limitations corresponding to the link organisation.

3.1 Estimation of available link capacity

As it was stated above, one can expect that capacity available for traffic in the access link is reduced comparing to the link bit-rate. In this section we will explain the factors causing this.

The formula for available capacity, C_{av}, of the access link with C kbps is the following:

$$C_{av} = C * \frac{N-M}{N} * a * r * b, \tag{1}$$

where:

a : is the ratio of the cell to slot time duration; this is called guard time. The value a is usually smaller than 1. Minimum value of parameter a depends on the assumed maximum distance between mobile terminal and the base station. For instance, a=0.7 for the link of 2048 kbps (El link) with GSM cell of 35 km. Exact value of a is tuned after measurements made in real system. Notice that for the indoor system a is closed to 1 while for outdoor systems a is much less than 1;

r: is the code rate resulting from application of redundancy code; for instance, using the Reed-Salomon code (71,55), r=55/71;

b: is the coefficient related to the assumed cell format in the access comparing to the standard; b=53/(53+2) for MEDIAN;

N: is number of slots in the frame; N=64 for the MEDIAN;

M: is number of control slots in the frame; M=3 for MEDIAN (for synchronisation, broadcast and contention slots);

For instance, C_{av}=728.5 kbps for the link with C=2048 kbps, a=0.5, r=55/71 and b=53/55. Notice that in this example the available capacity is reduced about 3 times (!) comparing to the link bit-rate. Anyway, only this capacity can be considered for serving traffic demanding guaranteed bandwidth in the wireless access link.

However, the estimated available link capacity is for two directions, up and down.

3.2 Conditions for the CAC

The discussed in this section issue corresponds to additional conditions for the MAC protocol for performing CAC function assuming that the wireless link capacity is evaluated by (1). Recall that CAC consists in refusing a new call if the addition of its traffic would lead to an unacceptable quality of service level for that or any previously accepted traffic.

Unfortunately, the implementation details of MAC protocol could introduce additional reducing of available bandwidth causing by: (1) incorrect setting of protocol parameters, (2) unpredictable protocol behaviour under high load conditions. Below, we show that the above limitations could lead to significant traffic service quality degradation. As example, we will examine the MEDIAN protocol, but the conclusions are more general.

3.2.1 Setting protocol parameters

In the MEDIAN protocol, it appears that Δ_{max} values assigned for each connection have essential impact on quality of carried traffic. Small Δ_{max} could cause cell loses because of allowed waiting time expiration while too large Δ_{max} usually leads to large CDV, non acceptable e.g. for CBR connections.

Let assume that N_{UP} running connections pass a RAP unit and, for the simplicity, all of them are of simplex type with up direction. The i-th connection, $i=1,\ldots,N_{UP}$, is characterised by its negotiated PCR(i) value. The following condition should be satisfied in order to avoid cell losses inside the connection i, caused by assumed TDMA frame structure:

$$\Delta_{\max}(i) \geq \left(4+\left(\sum_{k=1}^{N_{UP}}\left\lceil\frac{PCR(k)}{PCR(i)}\right\rceil-1\right)\right)*t_s \tag{2}$$

where t_s denotes time slot duration, $t_s = 53*8/(C*a*r*b)$.

Explanation of the formula (2) is the following. Under the worst case scenario, a cell from the i-th connection can wait for its transmission by the time required for three control slots (contention, synchronisation and broadcast, see Fig. 1) and for transferring cells belonging to the rest (N-1) running connections. Therefore, the number of such cells from the k-th connection with PCR(k) is $(PCR(k)/PCR(i))^+$.

Notice that (2) evaluates minimum CDV value the protocol guarantees.

3.2.2 Low load conditions

In this section we will argue that MEDIAN protocol can guarantee reasonable CDV values only in the case when it works under polling regime only (see section 2). For such conditions, the slots in the frame are assigned for given connection in a predefined pattern and are fixed during the connections are running. The cells waiting for up and down transmission are served in the dedicated slots only.

Consider the traffic carried by the system is generated along N_{UP} and N_{DOWN} connections, each of them is of CBR or VBR type. Each connection is characterised by its own PCR value.

The provided below analysis is derived from the point of view of the up connections. The condition that such system will work on the basis of polling scheme only is satisfied when maximum one cell from each connection can wait for its up transmission.

The number of slots L_{DOWN}, dedicated to the down connections, and transferred at the beginning of each MEDIAN frame is done by the formula:

$$L_{DOWN} = \sum_{i=1}^{N_{DOWN}} \left\lceil \frac{64 * t_S * PCR_i}{53 * 8} \right\rceil = \sum_{i=1}^{N_{DOWN}} \left\lceil \frac{64 * PCR_i}{C * a * r * b} \right\rceil, \tag{3}$$

where PCR_i is the peak cell rate value of the i-th connection, i=1,..., N_{DOWN}.

As a consequence, the low load conditions (serving according to the polling scheme only) are satisfy when:

$$\frac{53 * 8}{\max\limits_{i=1,\dots,N_{UP}} \{PCR_i\}} > t_s * (4 + (N_{UP} - 1) + L_{DOWN}), \tag{4}$$

Finally, the formula (5) gives the maximum number of connections, the system can support keeping low load conditions.

$$N_{UP} < \frac{C * a * r * b}{\max\limits_{i=1,\dots,N_{UP}} \{PCR_i\}} - 3 - L_{DOWN}, \tag{5}$$

For this case, the minimum CDV value (minimum Δ_{max}) the system can guarantee without cell loses is:

$$\Delta_{max} > \begin{cases} t_s * (4 + (N_{UP} - 1) + L_{DOWN}), & \textit{UP direction} \\ t_s * (64 + L_{DOWN} - 1), & \textit{DOWN direction} \end{cases} \tag{6}$$

Remark that number of connections the MEDIAN can support is conditioned by the maximum PCR value.

3.2.3 Unpredictable protocol behaviour under high load conditions

High load conditions in the wireless link could lead to cell transfer quality degradation. This is mainly caused by non-ideal slot reservation mechanism, which allows that some slots are transferred as empty despite that cells are waiting for transmission.

This is caused by the fact that more slot reservations than required are made by the protocol since a number of waiting cells is served by polling. As a consequence, such non-work conserving system behaviour leads to significantly larger CDV. Setting Δ_{max} for each connection according to (6) results that CLR essentially grows in the case of high load conditions.

4. NUMERICAL EXAMPLES

In this section we will show exemplary numerical results illustrating what one can expect about cell transfer characteristics when the RAP is controlled by MEDIAN protocol. These characteristics will be provided in terms of CDV and CLR parameters. More precisely, we will show the minimum CDV values satisfying that CLR=0. In the numerical examples we assumed that the link capacity C=2.048 kbps while C_{av}=728.5 kbps (a=0.5, r=55/71 and b=53/55).

The following types of connections were taken into account:

1. CBR connections:
   - CBR#1: with PCR=72,2 kbps (equivalent to 64 kbps, 8 kbps is for AAL1 and ATM headers [11]),
   - CBR#2: with PCR=18 kbps (equivalent to 16 kbps, 2 kbps is for AAL1 and ATM headers).
2. VBR connections:
   - VBR#1: ON/OFF source with PCR=72,2 kbps (with conditions as for CBR#1), mean burst duration: 350 msec, mean silence duration: 650 msec (VBR voice traffic model for PCM modulation technique),
   - VBR#2: ON/OFF source with PCR=18 kbps (with conditions as for CBR#2), mean burst duration: 350 msec, mean silence duration: 650 msec (VBR voice traffic model for LD-CELP modulation technique).

The presented below results correspond to two cases: (1) the cells are transmitted only to the up direction (from the terminals to the base station), (2) the cells transferred by the wireless link are transmitted to both up and down directions.

Case study no. 1: cells are transferred to the up direction only

The experiments were provided assuming that a number of identical types of connections are in progress, which are CBR#1, CBR#2, VBR#1 or VBR#2. The obtained simulation and theoretical (see formula (6)) results are reported in the tables 1-4. For each case, one can observe "low" and "high" traffic load conditions (see formula (4) and (5)). As it was expected, when the traffic is low then the CDV values are similar to these calculated by formula (6). On the other hand, high load conditions occurring on the considered link cause very large CDV that could be slightly accepted e.g. for CBR connections. Therefore, we stress that correct admission control with predictable CDV values is only possible under low load conditions. This is of course undesirable since the available bandwidth in the wireless access link is reduced again. For instance, the upper bound for low load conditions is:

- for CBR#1 connections: 7* 72,2 kbps = 505 kbps;
- for CBR#2 connections: 38* 18 kbps = 684 kbps;
- for VBR#1 connections: 7* 72,2 kbps *0.35 =177 kbps;
- for CBR#1 connections: 38* 18 kbps *0.35 = 240 kbps;

(in the analysed cases C_{av}=728.5 kbps).

On the basis of these results one can conclude as follows:

- Admission control with predictable CDV values is only possible on the basis of PCR declarations; therefore number of admitted connections does not depend on type of connection, CBR or VBR;
- Upper bound for keeping low load conditions can be determined on the basis of the PCR declarations;
- Effective bandwidth utilisation is lower in the case of VBR connections.

Table 1: CDV values for identical CBR connections with PCR=72,2 kbps, CLR=0

	Number of running CBR connections (UP direction)						
	1	3	5	7	8	9	10
	Low load				**High load**		
CDV [ms] Simulation results	**2.21**	**3.32**	**4.43**	**5.54**	**9.72**	**29.8**	**184.26**
CDV [ms] Theoretical results	**2.22**	**3.33**	**4.44**	**5.55**	-	-	-

Table 2: CDV values for identical CBR connections with PCR=18 kbps, CLR=0

	Number of running CBR connections (UP direction)										
	1	3	5	7	10	15	25	30	35	38	39
	Low load										High load
CDV [ms] Simulation results	**2.11**	**3.31**	**4.43**	**5.54**	**7.20**	**9.97**	**15.53**	**18.30**	**21.07**	**22.74**	**128.0**
CDV [ms] Theoretical results	**2.22**	**3.33**	**4.44**	**5.55**	**7.21**	**9.99**	**15.54**	**18.31**	**21.08**	**22.75**	**-**

Table 3: CDV values for identical VBR connections with PCR=72,2 kbps, CLR=0

	Number of running VBR connections (UP direction)									
	1	3	5	7	8	10	15	20	25	30
	Low load				High load					
CDV [ms] Simulation results	**2.20**	**3.32**	**4.43**	**5.54**	**7.21**	**29.0**	**215.0**	**927.0**	**3045.0**	**20000.0**
CDV [ms] Theoretical results	**2.22**	**3.33**	**4.44**	**5.55**	**-**	**-**	**-**	**-**	**-**	**-**

Table 4: CDV values for identical VBR connections with PCR=18 kbps, CLR=0

	Number of running VBR connections (UP direction)								
	1	**3**	**5**	**10**	**20**	**30**	**38**	**40**	**50**
	Low load							**High load**	
CDV [ms] Simulation results	**2.2**	**3.32**	**4.43**	**7.20**	**12.75**	**18.30**	**22.7**	**26.5**	**58.1**
CDV [ms] Theoretical results	**2.22**	**3.33**	**4.44**	**7.21**	**12.75**	**18.31**	**22.7**	**-**	**-**

Case study no. 2: cells are transferred to the up and to the down

Tables 5-6 show the exemplary results corresponding to the case of identical CBR (CBR#1 or CBR#2) connections with up and down directions. The reported CDV values are related to the up connections only. It appears that for the considered set of CBR#1 connections, the system behaves on the high load conditions (the formula (5) is not valid) only. Despite that one can observe small CDV values they are not predictable, what is very important for the CAC function. In the case of CBR#2 connections (see Table#6), low and high load conditions are clear. However, the conclusions are similar to these form the case study no. 1.

Table 5: CDV values for CBR bi-directional up connections with PCR=72.2, CLR=0

Number of running CBR connections (UP & DOWN direction)									
DOWN 72 kbps	**1**	**1**	**1**	**1**	**1**	**1**	**1**	**1**	**1**
UP 72 kbps	**1**	**2**	**3**	**4**	**5**	**6**	**7**	**8**	**9**
	High load								
CDV [ms] Simulation results	**5.57**	**6.65**	**7.76**	**8.98**	**10.25**	**14.38**	**20.83**	**30.73**	**2000**
CDV [ms] Theoretical results	**-**	**-**	**-**	**-**	**-**	**-**	**-**	**-**	**-**

Table 6: CDV values for CBR bi-directional up connections with PCR=18 kbps, CLR=0

Number of running CBR connections (UP & DOWN direction)									
DOWN 18 kbps	**1**	**1**	**1**	**1**	**1**	**1**	**1**	**1**	**1**
UP 18 kbps	**1**	**5**	**10**	**15**	**20**	**25**	**30**	**36**	**39**
	Low load								**High load**
CDV [ms] Simulation results	**3.32**	**5.54**	**8.31**	**11.09**	**13.8 6**	**16.64**	**19.41**	**22.74**	**1500**
CDV [ms] Theoretical results	**3.33**	**5.55**	**8.32**	**11.09**	**13.8**	**16.65**	**19.42**	**22.75**	**-**

Finally, tables 7 and 8 show the CDV values of CBR up connections for the case with a mix of CBR#1 and CBR#2. These results confirm the above conclusions. Similar experiments were provided for the VBR#1 and VBR#2 traffics.

Table 7: CDV values for CBR up connections with PCR=18 kbps and PCR=72.2, CLR=0

Number of running CBR connections (UP & DOWN direction)									
DOWN CBR#2 (18 kbps)	1	1	1	1	1	1	1	1	1
UP CBR#1 (72 kbps)	1	2	3	4	5	6	7	8	9
	Low load					**High load**			
CDV [ms] Simulation results	3.32	3.88	4.43	4.99	5.54	7.17	9.28	23.38	34.95
CDV [ms] Theoretical results	3.33	3.88	4.44	4.99	5.55	-	-	-	-

Table 8: CDV values for CBR up connections with PCR=18 kbps and PCR=72.2, CLR=0

Number of running CBR connections (UP & DOWN direction)									
DOWN (CBR#2) 18 kbps	2	2	2	2	2	2	2	2	2
UP CBR#1 (72 kbps)	1	2	3	4	5	6	7	8	9
	Low load			**High load**					
CDV [ms] Simulation results	4.43	4.98	5.54	6.1	7.63	8.87	17.95	25.74	42.14
CDV [ms] Theoretical results	4.44	4.99	5.55	-	-	-	-	-	-

5. SUMMARY

The paper evaluated ATM wireless access link controlled by MEDIAN protocol from the point of view of its ability for providing effective admission control. It appears that organisation of such link causes several limitations resulting that its available bit-rate capacity is significantly smaller than one can expect. The factors that have main impact on this are: (1) necessary overhead for providing appropriate frame structure, error protection mechanism and handling long distance terminals, (2) double slot reservation and polling mechanisms for assigning frame slots to active terminals. As a consequence, for the connections requiring rigorous QoS with respect to allowed maximum CDV, like e.g. these dedicated for voice transferring, the allowed traffic load in the RAP should be rather low. The presented exemplary numerical results show that for the case of high link utilisation CDV values are large and unpredictable. The upper bound of the link capacity available for performing CAC function can be evaluated by formulas presented in the paper.

REFERENCES

[1] B.Sklar „Rayleigh Fading Channels in Mobile Digital Communication Systems Part I: Characterization", IEEE Communication Magazine, September 1997

[2] B.Sklar „Rayleigh Fading Channels in Mobile Digital Communication Systems Part II: Mitigation", IEEE Communication Magazine, September 1997

[3] J.B. Cain, D. N. McGregor „A recommended error control architecture for ATM networks with wireless links", IEEE Journal on Selected Areas in Communications Vol. 15 No. 1 January 1997

[4] A. Bak, A.Beben, W.Burakowski, Z.Kopertowski, P.Pyda, M. Lesniewicz „Quality of ATM Layer Services in Wireless Networks" , Polish Teletraffic Symposium, Szklarska Poreba, Poland, 1999.

[5] ACTS Program Median develops a 60 GHz wireless LAN, (http://www.tno.nl/instit/fel/div3/median.html")

[6] ACTS 2004 Program SAMBA - System for Advanced Mobile Broadband Application (http://hostira.cet.pt/samba/Index.html)

[7] Frederic Bauchot, Stephane Decrauzat, Gerard Marmigcre, Lazaros Merakos, Nikos Passas , „MASCARA, a MAC protocol for wireless" (http://www.tik.ee.ethz.ch/~wand/DOCUMENTS/documents-frame.html)

[8] Markku Niemi, „Application Requirements for WATM", ATM Forum document no. 96-1058.

[9] J.Roberts, U. Mocci, J. Virtamo (Eds.), „Broadband Network Teletraffic. Performance Evaluation and Design of Broadband Multiservice Networks. Final Report of Action COST 242", 1996

[10] M. Noorchashm, B. Bharucha, and G. Wetzel, „Buffer Design for Constant Bit Rate Services in Presence of Cell Delay Variation" , ATM Forum document no. 95-1454.

[11] Recommendation ITU-T I.363, „B-ISDN ATM Adaptation Layer (AAL) Specification"

[12] R.R. Bhat, „Draft baseline text for Wireless ATM Capability Set 1 Specification", ATM Forum document BTD-WATM-01.10, December 1998

Minimum GPRS Bandwidth for Acceptable H.261 Video QoS

Iyad Al Khatib, Anders Franzen, Fabio Moioli
Department of Teleinformatics, The Royal Institute of Technology
Ericsson Business Networks, and Ericsson Wireless LAN Systems, Sweden
iyad@it.kth.se, {Anders.franzen, Fabio.moioli}@era.ericsson.se

Key words: GPRS, H.261 video codec, Hurst parameter, multiplexing gain

Abstract: As part of a larger research on multimedia traffic performance over GPRS, we present a QoS study focusing on one parameter: *bandwidth*. GPRS is an evolutionary phase and a critical step towards the third generation (3G) of mobile systems providing a data rate of 2Mbps. In our study on GPRS, we investigate *multimedia video traffic*. The video codec used is H.261 with QCIF resolution. Many parameters are under research; however, in this paper we focus on one parameter: *minimum required bandwidth for acceptable QoS of QCIF H.261 video streams over the wireless and mobile medium, GPRS.* Some other parameters of interest like the Hurst parameter and multiplexing gain are tackled.

1. INTRODUCTION

The wireless and mobile telecommunication world is experiencing a very critical transitional stage, where new QoS parameters are to be defined. Within a more general study on mobile systems evolution, we analyse multimedia traffic over GPRS, a new phase of mobile communication media. GPRS represents an evolutionary step from the existing GSM system, where its purpose is to bring packet switched data services to the mobile system. With GPRS, the user can always be connected to the network since charging is not based on the connection time. The final billing scheme is not totally defined yet, but the main point is that the user should not care aboutconnection time.

One of the other goals of GPRS is to try to provide higher speeds than traditional GSM systems. The maximum theoretical speed over GPRS is supposed to be around 115Kbps. This bandwidth is achieved with very good radio conditions, and when the network is fully developed. In practice, the starting GPRS speed would, to a large probability, be somewhere between 20Kbps and 56Kbps. An enhanced GPRS system called EDGE is supposed to bring the speed up to 384Kbps. This very evolutionary phase of mobile systems is believed to be only one step towards the third mobile systems generation (3G), which is expected to give speeds up to 2Mbps.

The GSM system uses Time-Division Multiple Access (TDMA) with eight radio frequency time slots. A network operator can dedicate 0 to 8 of these time slots to GPRS. Each mobile terminal can send/receive in 1 to 8 time slots. It is believed that the first mobile terminals generation for GPRS will support 4 time slots downlink and 1 time slot uplink, which gives around 14Kbps uplink and 56Kbps downlink.

With this great shift that GPRS will introduce to the wireless and mobile world, we are interested in investigating the quality of service that GPRS can offer to multimedia applications, mainly video quality. Our research in this area is long term; however, in this paper we investigate few multimedia traffic parameters for one video standard. The format of the video streams we investigate over GPRS is H.261 with QCIF resolution. H.261 is chosen for its low bit rate [10]. The H.261 video streams in the experiments are variable bit rate streams, which makes them more suitable for the medium [4]. Quarter-CIF (QCIF) has 176 pixels per line, and 144 lines [9]. QCIF is chosen, because it is mainly used for desktop videophone applications i.e. the size will be suitable for a mobile unit. In addition, all codecs must be able to handle QCIF.

The parameter we focus on throughout the experiments is the minimum GPRS bandwidth required for acceptable QoS of H.261 video streams of QCIF resolution. We are also interested in self similarity since if we can define which type of videos show self similarity over GPRS, then GPRS vendors can learn more about how to deal with video over this medium [3, 6]. In this respect, the Hurst parameter is calculated. The Hurst parameter can be looked at as a self-similarity value; if near to 1, then this would be a sign of self-similarity. However, if it shows a value nearer to 0.5, then there is not much of self-similarity in the traffic.

In a study of multimedia over GPRS, it is very important to note that the standards with which the QoS is judged are subjective. Unfortunately, up till now, the judgements on acceptable QoS for multimedia streams are relative to the observer's personal standards [8]. Hence, we find it very important that, in our study, the minimum acceptable parameters investigated are defined by a representative number of people from different population backgrounds. Hence a common acceptable QoS is set to find the minimum bandwidth sought.

To calculate the theoretical values for the minimum acceptable bandwidth, we use the Multiplexing Gain formula [5]:

$$G_n = nR_p/C_n \qquad \ldots (1)$$

where R_p is peak rate for the video stream; n is the number of independent streams combined for transmission; and C_n is the link-bandwidth required for the desired QoS for the multiplexed stream of n sources (C_1 being the link bandwidth for a single source).

$$(1) \Rightarrow G_n = nR_p/C_n = [nC_1/C_n][R_p/C_1] = [nC_1/C_n]G_1 \qquad \ldots (2)$$

where G_1 is the multiplexing gain for one source.

$$(2) \Leftrightarrow C_n = n[G_1/G_n]C_1 \qquad \ldots (3)$$

Here we think of the multiplexing gain as a parameter to use in order to achieve the minimum link-bandwidth for n streams where $n \in N^*$, the set of natural numbers - $\{0\}$. The multiplexing gain G_n for n number of independent streams is given by,

$$\frac{1}{G_n} = \frac{1}{b} + \left(\frac{1}{G_1} - \frac{1}{b}\right) n^{\frac{1-2H}{2H}} \qquad \ldots (4)$$

where b is the peak-to-average and H is the Hurst parameter. Many methods can be used to calculate the Hurst parameter, like time variance plot, R/S analysis [2], and periodogram method.

2. EXPERIMENTS AND RESULTS

Figure 1 shows the testbed, which consists of two video senders, a GPRS emulator, a receiver, and a traffic measurement and analysis tool, NIKSUN NetVCR™. All the experiments, except the last one, use one sender only, for they are dedicated to studying the bandwidth required for one video stream. On the other hand, the last experiment concentrates on the performance when multiple streams are sent over GPRS. Table 1 shows the video streams, where "Comm" is the stream used in experiments 1 to 5. The packet time slots on the GPRS medium are set to 8 time slots throughout all the experiments since using less for video transmission will not lead to acceptable QoS. First, we look into whether there is any difference between the behavior of two media: GPRS with no restricting limits, and 10-BT. The associated results for the H.261 video stream are presented in table 2, where one would conclude that when the GPRS is dedicated to one video stream, with no background users, it will most likely behave like Ethernet. However, when running the first experiment on 10 Mbps Ethernet, we got no missing frames at the receiver end, while in running the experiment over GPRS, with 12dB, we had 2,670 video frames missing out of 6,306 video frames of the same stream (see table 4, Exp. 1 and 2). Figures 1-8 show the number of bytes (vertical axis) versus the packet size categories (horizontal axis).

Table 1. Video sequences used in the experiments. "Comm" is used in experiments 1 to 5.

Type of video	Length (mm:ss)	Total bytes	Total packets	Average bandwidth (bps)	Hurst param.
Music 1	05:18	1,769,912	7,941	44,387	0.81
Music 2	03:39	3,027,270	4,745	109,584	0.78
Music 3	03:25	2,633,906	4,582	101,793	0.88
News	13:55	10,657,756	18,313	101,623	0.74
Talking head	12:51	11,682,038	13,682	120,901	0.61
Commercial **"Comm"**	05:06	3,631,412	6,942	94,322	0.79
Total	*44:14*	*33,402,294*	*56,205*		

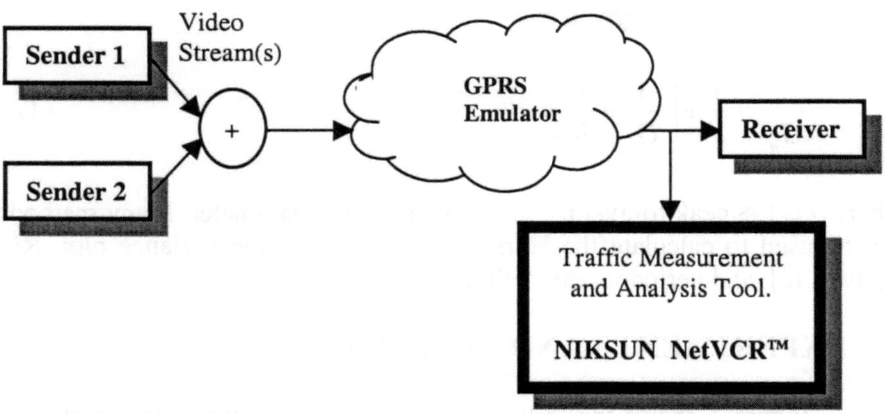

Figure 1. Testbed of bandwidth investigation for H.261 video quality over GPRS.

Hence in figure 1, the peak (bytes) is for the packets that are 512-1024 bytes in size, excluding 1024 byte packets. The second level peaks are for packets of 1024-2048 bytes (excluding 2048 byte packets) and 216-512 bytes (excluding 512 byte packets) respectively. Figures 2 through 8 can be read in a similar way for the associated experiments.

Table 2. Differences between GPRS with no restrictions and 10BT for "Comm". S/N=12 dB.

	No limits on GPRS	**10 Mbps Ethernet**
Total no. of bytes received	1.7284e+003	1.7289e+003
Median	3848	3788
Peak-to-Average ratio	5.3804	4.7394

We use the parameters in tables 3 and 4 to do some calculations for comparisons with the values received by the application. In this respect, we would like to note that the video is observed in real time and then the number of missing video frames is calculated. We believe that these numbers are very important to relate to acceptable QoS of the H.261 streams over GPRS. The fact that no frame is missing in experiment 1 can also be seen while watching the video stream live. Experiment 1 is also used as a comparison basis for acceptable QoS.

Table 3. Number of packets vs packet size for "Comm" video stream.

Packet Size Categories (Bytes)	**Count (Packets)**				
	10 BT	*GPRS*			
	Exp.1,fig.1	**Exp.2**,*fig.2* 12dB 0 BGU	**Exp. 3**,*fig.3* 15dB 0 BGU	**Exp. 4**,*fig.4* 12dB 20 BGU	**Exp.5**,*fig.5* 12dB 40 BGU
0 to 128	507	305	271	282	275
128 to 256	1089	651	594	585	623
256 to 512	2332	1332	1365	1302	1367
512 to 1024	2152	1312	1303	1232	1302
1024 to 2048	862	658	672	606	641
2048	0	14	19	12	21

Table 4. Statistics for "Comm" Video Stream over 10 Base-T and GPRS.

	10 BT	GPRS			
	Exp. 1	**Exp .2** 12dB 0 BGU	**Exp. 3** 15dB 0 BGU	**Exp. 4** 12dB 20 BGU	**Exp. 5** 12dB 40 BGU
Total Number of Bytes	7262824	4639074	4674424	4366946	4614556
Average Rate (bps)	48418.83	77317.90	103876.09	72782.43	102545.69
Number of Packets	13884	8544	8448	8038	8458
Average Rate (pps)	11.57	17.80	23.47	16.75	23.49
Minimum Packet Size	67	69	68	67	69
Maximum Packet Size	1066	1066	1066	1066	1066
Mean Packet Size	523.11	542.96	553.32	543.29	545.58
Packet Size Variance, B^2	93519.15	101585.97	100581.10	99668.34	99399.37
Variance/Mean	178.78	187.10	181.78	183.45	182.19
Missing Video Frames	0	2670	2719	2811	2788

In the second experiment GPRS is used. The S/N is 12dB i.e. worst case. However the main concern of this experiment is to measure the same parameters as in experiment 1 but over GPRS, hence we have no BackGround Users (BGU) i.e. the "Comm" video has all the bandwidth. The results are presented in figure 2. Collecting these parameters, we can also find a great similarity in the QoS delivered as well as the shapes of the graphs in figures 1 and 2. Around 41% of the frames are missing, but still the QoS is acceptable. The receiver end shows a rate ranging between 48Kbps and 62Kbps, which is a very good rate in our point of view for a video transmission with a QCIF resolution.

We also investigate the behavior and the used-bandwidth results for the 15dB S/N. We run exactly the same experiment as in experiment 2, but with 15dB instead of 12dB. The result is shown in figure 3. This experiment shows just a slight difference where there are more of large packets i.e. the traffic concentration is more in the middle (512-1024B) than in experiment 2. The rate ranges between 48Kbps and 63Kbps i.e. acceptable.

To get more practical results, we force background users over the GPRS network. The number of background users we have in experiment 3 counts to 20, with 12dB. The result can be seen in figure 4. The behavior still shows a graph similar to the previous experiments. The rate at the receiver's end still shows a range between 40Kbps and 60Kbps.

In the fifth experiment we force 40 users with 12dB over GPRS. The behavior is similar to the previous experiments in terms of graphical shape (figure 5), but the video quality drops down. In fact it is not acceptable at all. However, the rate at the receiver's end still shows a range between 48Kbps and 60Kbps.

In the sixth experiment our concentration is on the bandwidth when multiple streams are injected over GPRS, 12dB and 40 BGU. The results are predictable as shown in figures 6, 7 and 8. Filtering the traffic of each stream alone is important to study the bandwidth from the multiplexing gain point of view. Figure 6 shows the result for the traffic of both H.261 video streams at the same time. Since both streams, when injected together, have their first level peaks at (512-1024B), as well as their second level peaks at the same points (figures 7 and 8), then adding the two would lead to a graph with peaks at the same relative points (figure 6). For the first stream, the rate at the receiver's end still shows a range between 21Kbps and 37Kbps. For the second stream, the rate at the receiver's end still shows a range between 30Kbps and 50Kbps. The rates show that the bandwidth is divided, and this is a normal behavior. Around 60% of the frames are lost in each video stream. The visual effects on the QoS can be observed, and the delay between the frames is not within the acceptable range when there are transitions in the video stream. We also run multiple streams over GPRS to investigate more on the multiplexing gain and suitable bandwidth for the set QoS. Results are shown in table 7.

Referring to equation 3, if we know G_1 and C_1, then knowing C_n will be just a matter of knowing G_n, which can be calculated using equation 4. For an acceptable QoS, we will use equation 1 to get to a $C_1 = R_p$, where R_p is investigated in the experiments to be around 70Kbps. This makes $G_1 = 1$ for acceptable QoS. Hence (3) becomes: $C_n = n(1/G_n)(70) = 70n(1/G_n)$.

Table 5. Number of packets vs packet size; 2 video streams; GPRS, 12dB, 40 BGU.

Packet Size Categories (Bytes)	**Count (Packets)**		
	GPRS, 12 dB, 40 BGU		
	Total of Two video streams	First video stream **filtered**	Second video stream **filtered**
0 to 64	355	-	-
64 to 128	307	162	144
128 to 256	824	366	313
256 to 512	1433	747	699
512 to 1024	1452	730	728
1024 to 2048	624	304	305
2048	13	0	8

Table 6. Two video streams over GPRS, 12dB, 40 BGU.

	GPRS, 12dB, 40 BGU		
	Total of Two video streams	First video stream **filtered**	Second video stream **filtered**
Total number of Bytes	4949190	2456798	2401096
Average Rate (bps)	94270.27	16378.65	45735.16
Number of Packets	10016	4618	4394
Average Rate (pps)	23.85	3.85	10.46
Minimum Packet Size	60	72	68
Maximum packet size	1066	1066	1066
Mean Packet size	494.13	532.00	546.45
Packet Size Variance, B^2	10533.95	96162.32	97318.30
Variance/Mean	213.17	180.75	178.09
Missing video frames		4499	4399

Table 7. Multiplexing gain and min. bandwidth for a increasing number of video streams.

No. of Streams	Average (bits/interval)	Peak (bits)	Peak-to-Average	Hurst Param.	Mux Gain	=>	**Minimum Bandwidth**
2	198,307	278,528	1.40	0.87	1.82	=>	76.9Kbps
3	297,458	455,384	1.53	0.79	1.83	=>	114.8Kbps
5	487,719	628,664	1.29	0.82	1.60	=>	218.8Kbps
10	970,794	1,246,264	1.28	0.83	1.50	=>	466.7Kbps
15	1,455,795	2,002,776	1.38	0.84	1.54	=>	681.8Kbps

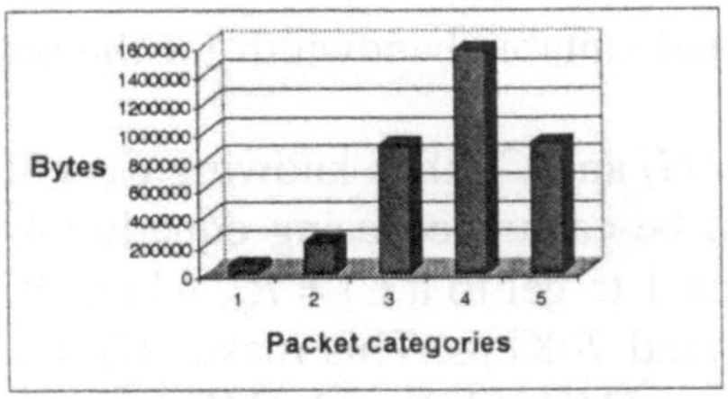

Figure 1. Exp. 1, "Comm" bytes vs packet categories over 10-BT.

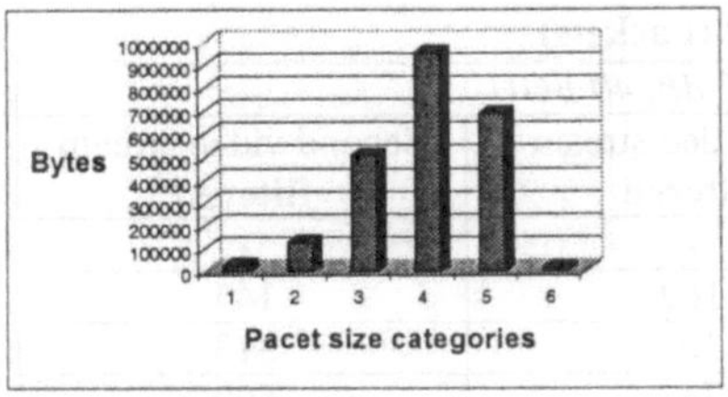

Figure 2. Exp. 2, "Comm" bytes vs packet size categories; GPRS, 12dB, 0 BGU.

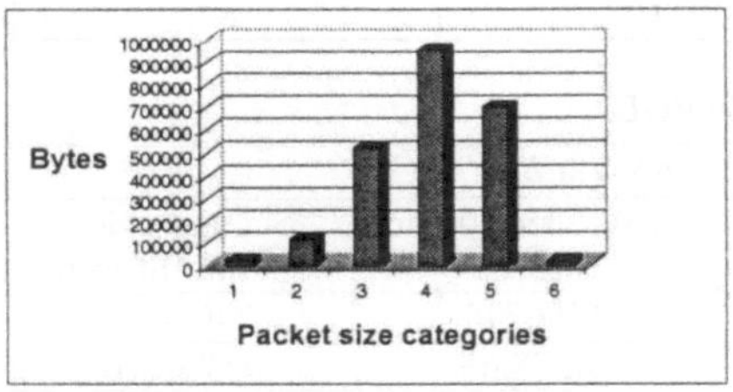

Figure 3. Exp. 3, "Comm" bytes vs packet size categories; GPRS, 15dB, 0 BGU.

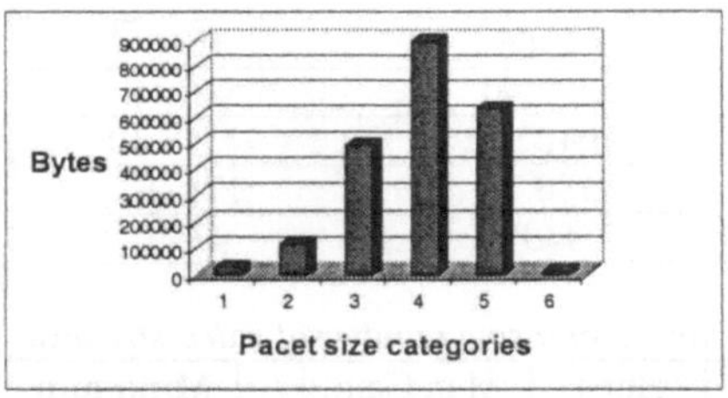

Figure 4. Exp. 4, "Comm" bytes vs packet size categories; GPRS, 12dB, 20 BGU.

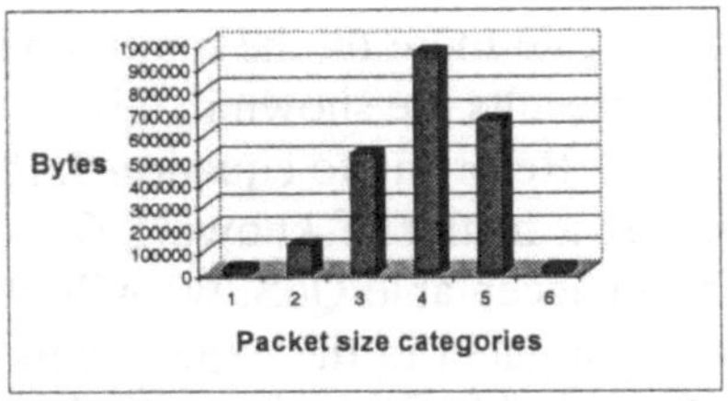

Figure 5. Exp. 5, "Comm" bytes vs packet size categories; GPRS, 12dB, 40 BGU.

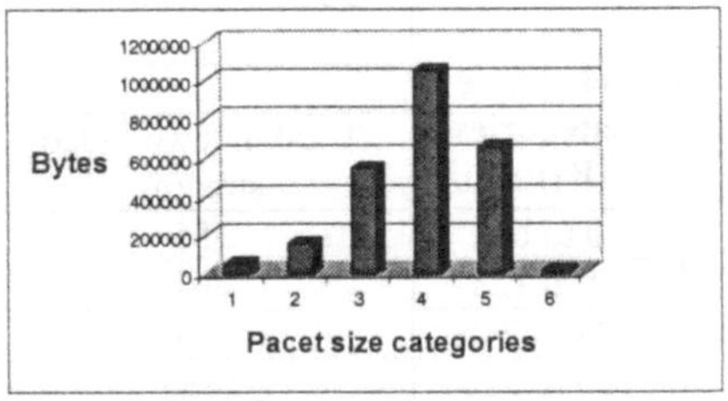

Figure 6. Exp. 6, two video streams traffic over GPRS, 12dB, 40 BGU.

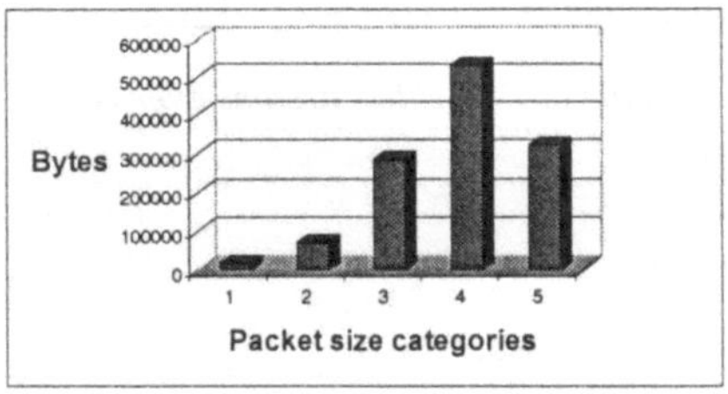

Figure 7. Exp. 6, first video stream over GPRS, 12dB, 40 BGU.

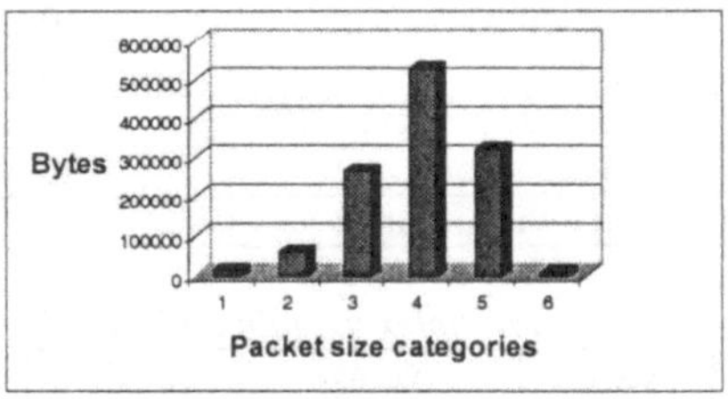

Figure 8. Exp. 6, second video stream over GPRS, 12dB, 40 BGU.

Figures 1-8. Horizontal axes show packet size categories of size X bytes, where labels 1, 2, 3, 4, 5, and 6 represent categories with 1 ≤ X < 128 bytes, 128 ≤ X < 256 bytes, 256 ≤ X < 512 bytes, 512 ≤ X < 1024 bytes, 1024 ≤ X < 2048 bytes, and X = 2048 bytes respectively.

3. EVALUATION OF RESULTS

As we increase the number of users and limitations on the GPRS, our calculations lead to a final value that we would like to present. The minimum acceptable bandwidth for H.261 video streams QoS over GPRS is found - after many iterations and trials - to be around 70-80Kbps for one QCIF H.261 video stream. This number is not very satisfactory since the practical limit that GPRS can deliver now is around 50Kbps. However, work is going on to reach higher practical limits, and if 70Kbps is reached, then sending video streams with QCIF resolution will be possible for the defined QoS. When a bandwidth of less than 70Kbps over GPRS is reached, the video quality and the missing frames number are not acceptable. In this respect, and regarding the transmission of multiple streams to one receiver to two different application port numbers, the sharing of the bandwidth will surely happen. However, the results in table 7 clearly show that the bandwidth needed over GPRS for the H.261 video for (n) streams will be less than the sum of the peak rates of the two streams. In other words, multiplexing gain will occur and will be a value greater than 1;

$$C_n = nR_p/G_n < nR_p = nC_1\,; \qquad G_n > 1, n \in N^*.$$

The quality with multiple streams will always be less than for one video sent as shown in experiment 6.

One parameter that seems promising for more research is the Hurst parameter shown in figure 9 with a *Log variance* vs *Log lag* plot. Since the self similarity is an interesting parameter to look at when all the presented data is available [7], we look at the Hurst parameter for two streams. The two H.261 video streams over GPRS show a Hurst parameter of around 0.97, with 12dB, 40 BGU. This means that the self-similarity is highly probable to occur [1]. This still needs more study to be conducted, but it is a very interesting start.

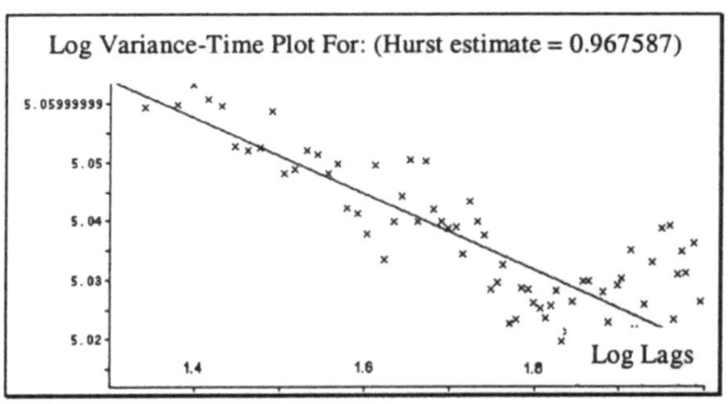

Figure 9. Hurst parameter plot for 2 video streams over GPRS, 12dB and 40 BGU.

4. CONCLUSION

We investigate some multimedia traffic parameters over GPRS, the third generation of mobile systems. The video streams investigated are encoded in H.261 codec. QCIF resolution is chosen for investigation since it can be deployed on mobile units. The minimum bandwidth required for acceptable QoS of QCIF H.261 video is dependent on the peak rate of the video, the number of streams, and how much the medium can have of multiplexing gain. For one video stream, the minimum bandwidth is around 70Kbps, which is still not easy to achieve over GPRS. However future GPRS generations will be able to supply this bandwidth and more. The encouraging part is that when two or more video streams are injected, they need less bandwidth than the sum of the peak rates of each. We hope that our study triggers more investigation in the filed of multimedia over GPRS from the traffic analysis point of view. The Hurst parameter is also presented briefly.

REFERENCES

[1] Jan Beran, Robert Sherman, Murad S. Taqqu and Walter Willinger, "Long-Range Dependence in Variable-Bit-Rate Video Traffic", IEEE Transactions on Communications 43, No. 2/3/4, pp. 1566-1579, 1995.

[2] M. W. Garrett, W. Willinger, "Analysis, Modeling and Generation of Self-Similar VBR Video Traffic", Computer Communications review vol. 24, no. 4, pp. 269-280, 1997.

[3] Daniel P. Heyman and T. V. Lakshman, "What Are the Implications of Long-Range Dependence for VBR-Video Traffic Engineering?", IEEE/ACM Transactions on Networking, vol. 4, no. 3, June 1996.

[4] C. Huang, M. Devetsikiotis, I. Lambadaris and A.R. Kaye, "Modelling and Simulation of Self-Similar Variable Bit Rate Compressed Video: A Unified Approach", Proc. of the ACM Sigcomm'95, Boston, pp. 114-125, 1995.

[5] K. R. Krishnan, A.L. Neidhardt, and A. Erramilli, "Scaling Analysis in Traffic Management of Self similar Processes", ITC 15, Elsevier Science B.V., 1997.

[6] W. Leland, M. Taqqu, W. Willinger and D. Wilson, "On the Self-Similar Nature of Ethernet Traffic", IEEE/ACM Transactions on Networking vol. 2, no. 1, February 1994.

[7] Kihong Park, Gitae Kim Mark Crovella, "On the Effect of Traffic Self-similarity on Network Performance", Proceedings of the 1997 SPIE International Conference on Performance and Control of Network Systems.

[8] P. Pruthi, D Ilie, A. Popescu "Application Level Performance of Multimedia Services", SPIE International Symposium on Voice, Video, and Data Communications Boston, Sep 19-22 1999.

[9] Thierry Turletti, "H.261 Software Codec for Videoconferencing over the Internet", Rapports de Recherche, Unité de recherche no. 1834, INRIA-SOHIA Antipolis, 1993.

[10] Draft ITU-T Recommendation H.263 (1996): "Video Coding For Low Bitrate Communication".

A Performance Analysis of IEEE 802.11 Networks in the Presence of Hidden Stations

Marek Natkaniec, Andrzej R. Pach
University of Mining and Metallurgy, Department of Telecommunications, Cracow, Poland
natkanie@kt.agh.edu.pl, pach@kt.agh.edu.pl

Key words: Wireless LAN, IEEE 802.11, Hidden Stations, Performance Analysis

Abstract: IEEE 802.11 is a wireless network standard that was completed in 1997. Unfortunately, the medium access protocol described in the standard meets some problems that arise from the presence of so-called hidden stations. This situation can cause degradation of the network performance. The paper describes a simulation analysis of influence of hidden stations on the IEEE 802.11 network efficiency in four different hidden terminals scenarios. The throughput and the mean packet delay as a function of the offered load has been studied. The presented results allow us to determine the usefulness of RTS/CTS mechanism usage in the presence of hidden stations.

1. INTRODUCTION

We can observe a permanent growth of interest in the area of WLANs (Wireless Local Area Networks) in last years. WLANs assure easy and free access to existing network infrastructures: LAN, MAN and WAN. These networks can also be attractive for new users by assuring wireless access to databases in magazines, stores, hospitals, airfields, museums etc. IEEE 802.11 [4] is a new wireless network standard that was completed in 1997. There are some vendors of IEEE 802.11 network cards.

The most common medium access algorithm used in WLANs is CSMA (Carrier Sense Multiple Access). In CSMA, every contending station senses the carrier before the transmission. Carrier sense allows avoiding the collisions by testing the signal energy in the occupied band. WLANs use a mutation of that scheme called CSMA/CA (Carrier Sense Multiple Access with Collision Avoidance). This algorithm has been employed by the DCF (Distributed Coordination Function) function of IEEE 802.11.

The basic idea of operation of protocols based on CSMA relies on packet transmission avoidance by the station if it detects that the radio channel is busy. Unfortunately, this class of protocols meets some problems that arise from the presence of hidden stations. The situation when some stations do not hear each other happens very often in WLANs. The station sending a data packet cannot be sure whether the packet reaches the destination without collision. This can happen, for example, when a station is not able to hear some other transmissions directed to the same receiver (some stations can operate at geographically separated areas). This situation can cause a substantial degradation of the network performance. The hidden, exposed and intruding station problems are described in details in the next section.

The first protocol that limits an unprofitable hidden stations influence in a single channel has been proposed in [5]. It was called MACA (Multiple Access Collision Avoidance). This protocol use two-way handshaking between the source and destination station. The source station transmits the RTS (Request To Send) packet to the destination station. If it receives the RTS packet correctly, then immediately starts to transmit the mini-packet called CTS (Clear To Send). The proper reception of the CTS packet means that the medium was reserved and the source station can start to transmit its data. There are many other protocols based on RTS/CTS exchange before the data transmission that operates in a single channel [2], [3], [4]. The DFWMAC protocol that realizes the DCF function of the IEEE 802.11 standard belongs to this group.

An influence of hidden stations on fairness of operation and performance of an IEEE 802.11 network has intensively been studied in the literature. An analysis of RTS/CTS usage in DFWMAC protocol was presented in [10]. This work presents the simulation results of a network consisted of eight stations with three hidden ones. It shows the positive effect of RTS/CTS usage. An analysis of fairness of a network consisted of 25 stations with four hidden ones was described in [11]. This work evaluates the network efficiency while exchanging different traffic types generated by different applications with and without usage of RTS/CTS mechanism.

This paper describes a simulation analysis of hidden station influence on the IEEE 802.11 network efficiency for four different hidden terminals scenarios. The throughput (overall and obtained for every station) and the mean packet delay as a function of the offered load has been studied. The presented results allow us to determine the usefulness of RTS/CTS mechanism usage in the presence of hidden stations.

2. HIDDEN, EXPOSED AND INTRUDING STATIONS

The limited area of transmission is characteristic for station operation in a wireless environment. It means that packets transmitted over wireless medium can be received by the stations which are located in the coverage area of the sender. The situation where all stations hear each other occurs very rarely. It brings the problem of hidden stations. A very similar problem arises in the case of exposed stations, that is described below. The presence of any of the mentioned cases brings a serious problem concerned with fair access. It causes the degradation of the network performance. There are four cases of hidden and exposed stations: a hidden sender, a hidden receiver, an exposed sender and an exposed receiver [1], [2].

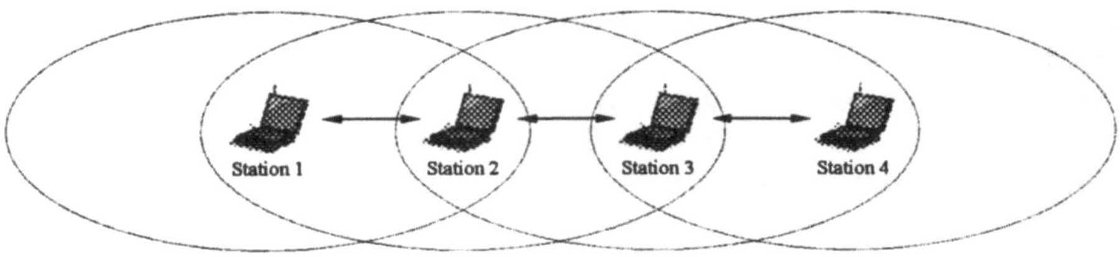

Fig. 1. En example of hidden and exposed stations

Fig. 1 presents four stations that are located in such a way that each station can hear the transmission from the immediate neighbors. When Station 1 transmits a packet to Station 2, Station 3 does not hears the transmission form Station 1. It could happen that during this time Station 3 transmits a packet, so the collision occurs. The Station 2 hears the collision but Station 1 does not. In this situation, when Station 1 transmits to Station 2, Station 3 should defer its transmission. This is the problem of hidden sender. The Station 2 should notify Station 3 about the transmission because Station 1 is not able to do it. Then, every data packet should be preceded by a mini packet handshake. A station should defer its transmission when in response it hears the handshake mini packet. From the above, it follows that Station 3 should defer transmission up to the end of data transmission from Station 1 to Station 2.

The next case is the situation where Station 4 wants to transmit a data packet to Station 3. At first, it sends the control mini packet to Station 3. Station 3 cannot send any packet because it defers. Station 4 will not hear a response then try to retransmit the control packet in order to establish the handshake. This is the case of a hidden receiver.

The case of an exposed station occurs when the station is within the range of the transmitter and out of range of the receiver. The problem of an exposed sender appears when the exposed station cannot start its own transmission during any other one because it cannot hear the response of a mini packet in the handshake. Let us consider again the case presented in Fig. 1. Station 3 transmits the packet to Station 4. The Station 2 cannot start its transmission to Station 1 because this can cause the collision of packets at Stations 2 and 3.

The exposed receiver problem can be explained as follows. Station 3 transmits a packet to Station 4. If Station 1 transmits a control mini packet to Station 2, Station 2 is not able to understand any transmission because of collision with transmission from Station 3.

It could also be a situation where the problem of an intruding terminal occurs. This is concerned with mobility of stations. When a terminal moves into the communication range of an occupied receiver, any transmission of the intruding station will cause the collision with any ongoing transmission. The protocols based on single channel environment are endangering to the intruding terminal problem. An unprofitable influence of intruding station can be reduced with the aid of multichannels protocols [3], [7].

3. OPERATION OF RTS/CTS MECHANISM

The IEEE 802.11 standard supports two access methods: a mandatory Distributed Coordination Function (DCF) method which is available in both ad hoc and infrastructure configurations, and an optional Point-Coordinated Function (PCF) which is available in certain infrastructure environments and can provide time-bounded services [6].

DCF is the fundamental access method used to support asynchronous data transfer on the best effort basis. All the stations must support DCF. DCF employs the carrier sensing (CS) mechanism. In order to minimize the probability of collisions a random backoff mechanism is used to randomize moments at which medium is tried to be accessed.

The DCF protocol is enhanced further by provision of a virtual CS indication called Net Allocation Vector (NAV) which is based on duration information transferred in special RTS/CTS frames before the data exchange. It allows stations to avoid transmission in time intervals in which the medium is surely busy. The detailed DCF description can be found in [4].

The handshaking usage allows increasing the network performance. The collisions of short information packets and reduction of an unprofitable hidden stations influence can increase the throughput.

4. RESULTS OF SIMULATION

The carried out simulations allows us to determine the realized throughput and the mean packet delay versus the offered load for each station while transmitting 1000 octets data frames for four different hidden station scenarios that are depicted in Fig. 2. The RTS/CTS mechanism was always enabled for all the scenarios in the first part of the study. The possibility of handshaking was disabled for Scenario C and D in second part of our study. The dependencies between the overall throughput and the mean packet delay (for all stations) versus the offered load for all considered scenarios are presented in figures in the following.

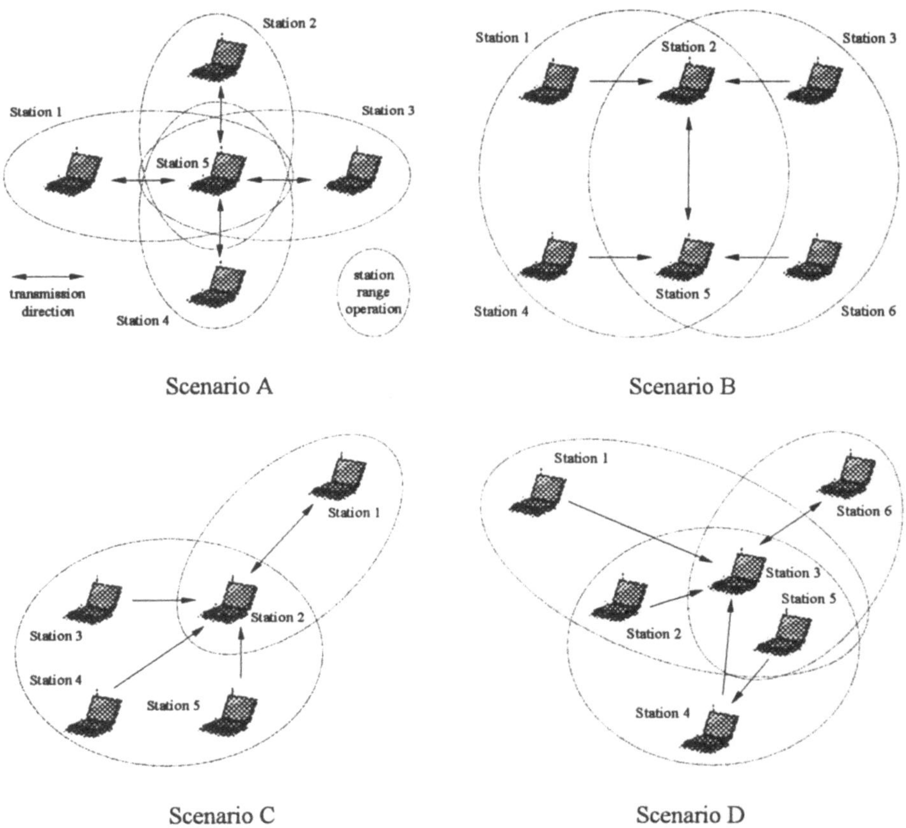

Fig. 2. Four examples of hidden station scenarios

The packet arrivals were realized according to the Poisson distribution. Several assumptions were made to reduce the complexity of the simulation model:

- The channel is error-free that means that each packet that is transmitted by the sender is successfully received by the receiver.
- The effects of propagation delay are neglected. This is very realistic assumption if the distances are of tens meters between stations.
- There is no station operating in the "power-saving" mode. Each station is "awake" all the time. Then transmitted frames can be received immediately by the destination stations.

The RTS/CTS/DATA/ACK or DATA/ACK mode of transmission was assumed. The network was configured to 2 Mbps medium capacity. Almost all parameters were taken from the standard specification and are adequate to the FHSS (*Frequency Hopping Spread Spectrum*) physical layer specification. The parameters used throughout all simulations are displayed in Table 1.

Table1 Parameters used throughout all simulations.

Parameter	Value	Parameter	Value
SIFS	28 μs	Minimum number of slots – CWmin	32 slots
DIFS	130 μs	Maximum number of slots – CWmax	1024 slots
Length of RTS	20 octets	Physical layer preamble	18 octets
Length of CTS	14 octets	Medium capacity	2 Mbps
Length of ACK	14 octets	Length of DATA frames	1000 octets
DATA header	32 octets	Number of retransmissions of RTS frames	4
Slot time	50 μs	Number of retransmissions of DATA frames	4
T1 timer	300 μs	Number of hidden stations	variable
T3 timer	300 μs	Number of stations	variable
Buffer size	10 frames	RTS_Threshold	RTS/CTS - enabled or disabled

The results of simulations are presented in a number of plots in the following:

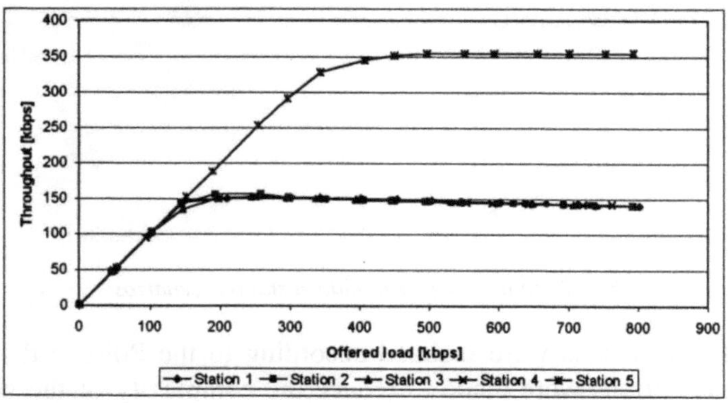

Fig. 3. Throughput versus offered load for Scenario A, RTS/CTS mechanism always enabled

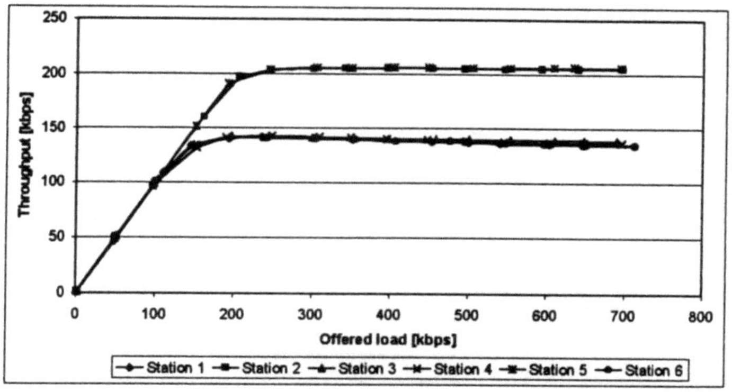

Fig. 4. Throughput versus offered load for Scenario B, RTS/CTS mechanism always enabled

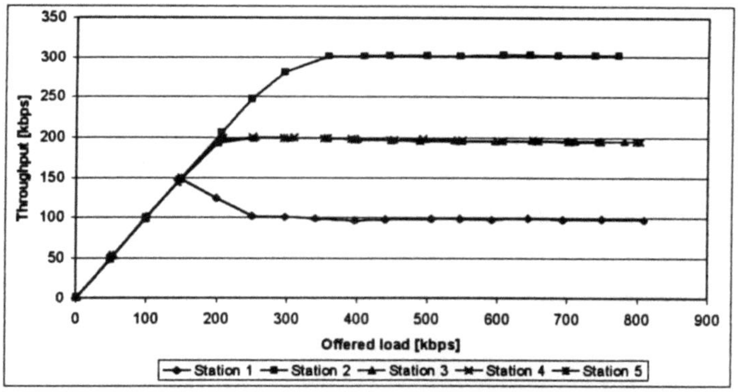

Fig. 5. Throughput versus offered load for Scenario C, RTS/CTS mechanism always enabled

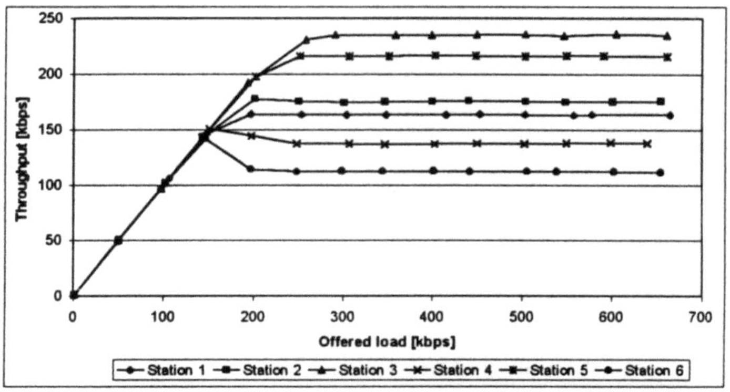

Fig. 6. Throughput versus offered load for Scenario D, RTS/CTS mechanism always enabled

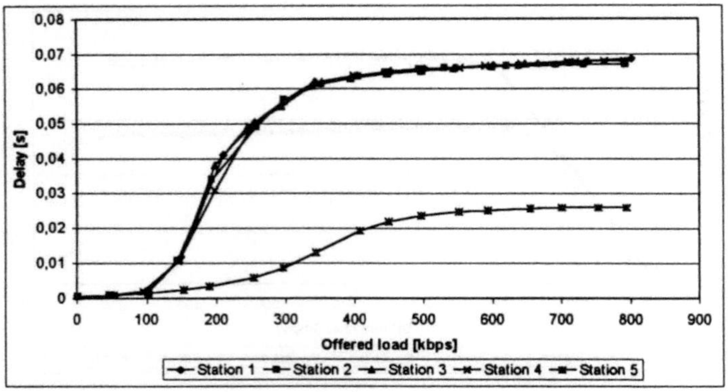

Fig. 7. Mean packet delay versus offered load for Scenario A, RTS/CTS mechanism enabled

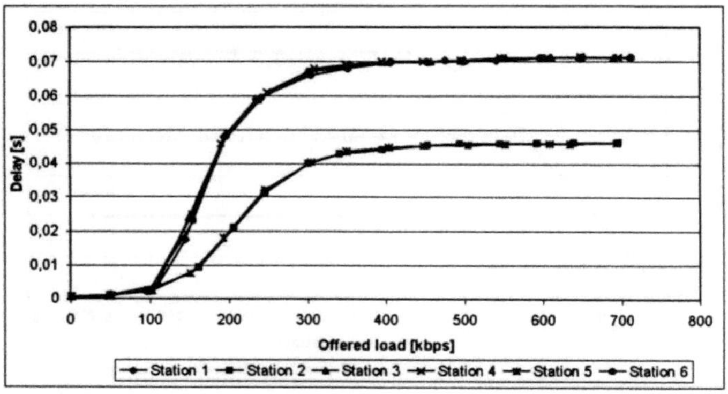

Fig. 8. Mean packet delay versus offered load for Scenario B, RTS/CTS mechanism enabled

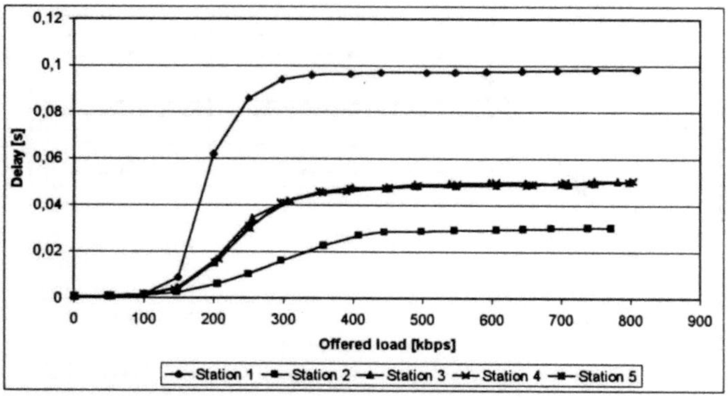

Fig. 9. Mean packet delay versus offered load for Scenario C, RTS/CTS mechanism enabled

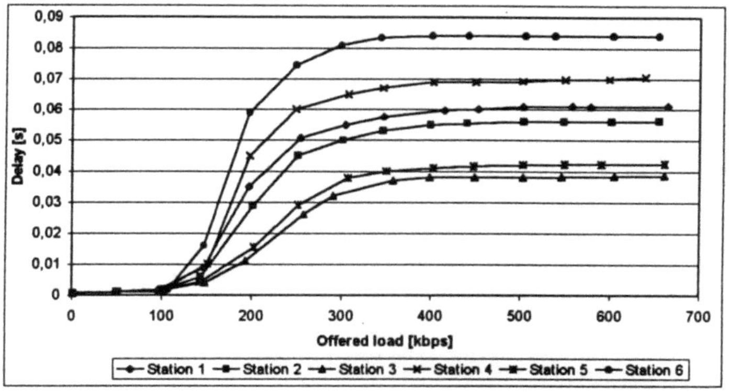

Fig. 10. Mean packet delay versus offered load for Scenario D, RTS/CTS mechanism enabled

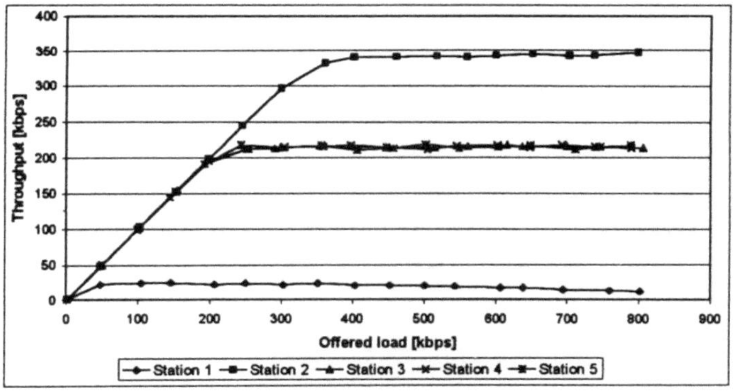

Fig. 11. Throughput versus offered load for Scenario C, RTS/CTS mechanism always disabled

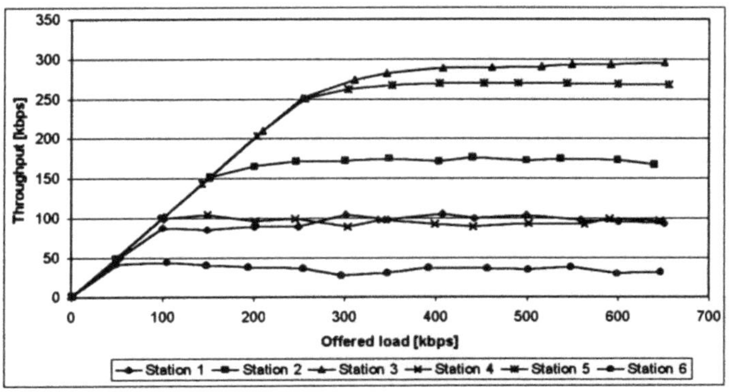

Fig. 12. Throughput versus offered load for Scenario D, RTS/CTS mechanism disabled

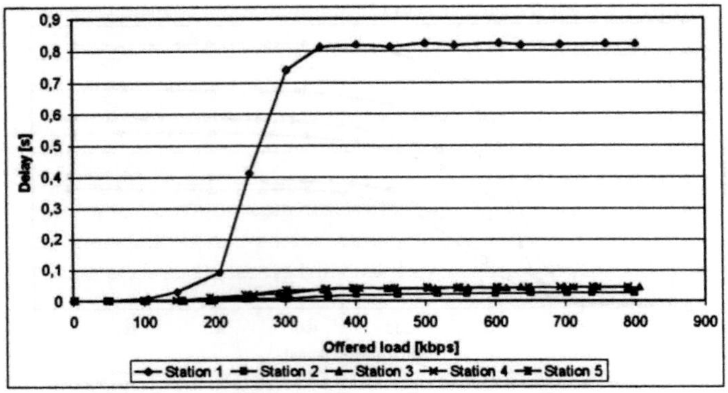

Fig. 13. Mean packet delay versus offered load for Scenario C, RTS/CTS mechanism disabled

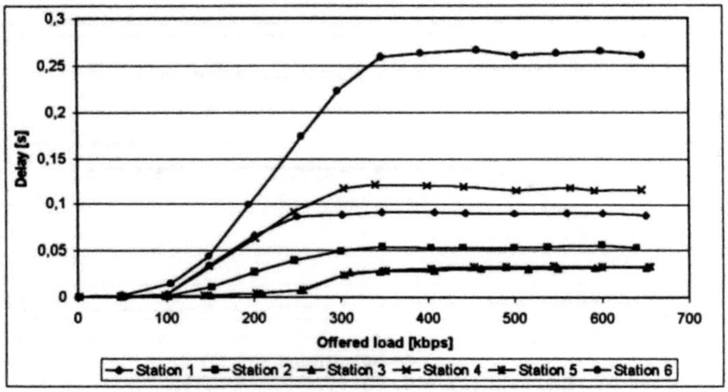

Fig. 14. Mean packet delay versus offered load for Scenario D, RTS/CTS mechanism disabled

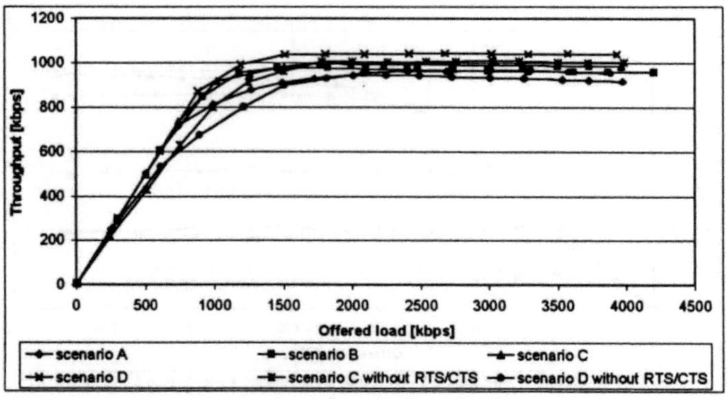

Fig. 15. Throughput versus offered load for all 6 scenarios

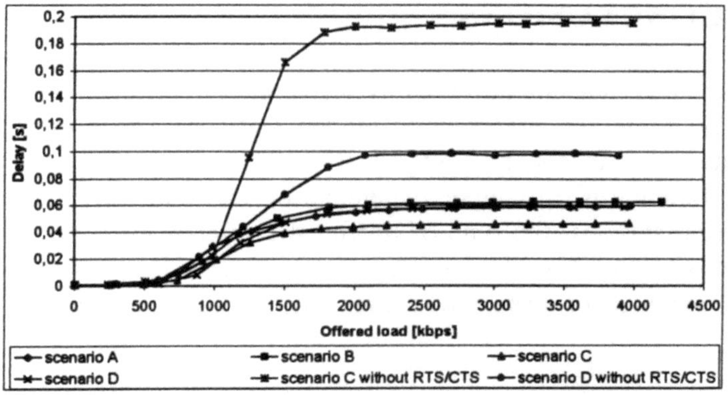

Fig. 16. Mean packet delay versus offered load for all 6 scenarios

5. DISCUSSION

The presented work describes a simulation analysis of hidden station influence on the IEEE 802.11 network efficiency. Four different hidden station scenarios were investigated. The throughput (overall and obtained for every station) and the mean packet delay as a function of the offered load were studied. They allow us to determine the usefulness of RTS/CTS mechanism usage. The obtained results allow us to draw some general conclusions about the IEEE 802.11 network efficiency in the presence of hidden stations:

- The presence of hidden stations brings significant network performance degradation.
- A growth of the offered load above the nominal capacity of the network does not brings the degradation of the realized throughput like in some others wireless networks (IEEE 802.11 MAC protocol is more stable).
- The presence of hidden station causes unfairness in access to the medium – stations located in the center have larger transmission privilege.
- Hidden stations frequently loose medium access competition for larger values of the offered load and, therefore, reach smaller throughputs and considerable higher delays.
- The RTS/CTS mechanism improves the fairness of network operation. It brings the growth of the realized throughput (even of hundreds percents) and reduction of the mean packet delay (up to many times) for hidden stations.
- The positive influence of the RTS/CTS mechanism usage grows with the number of stations (especially hidden), offered load and length of transmitted packets.

6. CONCLUSIONS

The presented analysis allow us to determine the efficiency of an IEEE 802.11 network in the presence of hidden stations. The presented study demonstrates the reasonableness of RTS/CTS mechanism usage. The obtained efficiency is, of course, dependent on many factors as the number of contending stations, the type of traffic, the offered load, the number of hidden stations, etc. The obtained results shows that it is better to use the RTS/CTS mechanism especially when the hidden stations location is unknown.

REFERENCES

[1] BHARGHAVAN V., *Performance Analysis of a Medium Access Protocol for Wireless Packet Networks.* IEEE Performance and Dependability Symposium '98, Raleigh, NC. August 1998.

[2] FULLMER C. L., GARCIA-LUNA ACEVES J. J., *Solution to Hidden Terminal Problems in Wireless Networks.* ACM SIGCOMM Conference on Communications Architectures, Protocols and Architectures, Cannes, France, September 1997

[3] HAAS Z. J., *On the performance of a medium access control scheme for the reconfigurable wireless networks.* IEEE Milcom'97, Nov. 1997

[4] IEEE 802.11 *Standard for Wireless LAN: Medium Access Control (MAC) and Physical Layer (PHY) Specification.* New York, IEEE Inc. 1997

[5] KARN P., *MACA – A new channel access method for packet radio.* in ARRL/CRRL Amateur Radio 9th Computer Networking Conference 1990

[6] NATKANIEC M., PACH A. R: *Simulation Analysis of Multimedia Streams Transmission in IEEE 802.11 Networks.* – ISWC'99 IEEE International Symposium on Wireless Communications, June 1999, Victoria, Canada

[7] SOUSA E., SILVESTER J. A., *Spreading code protocols for distributed spread-spectrum packet radio networks.* IEEE Trans. Commun., vol. 36 no. 3, Mar. 1988

[8] TOBAGI F. A., KLEINROCK L., *Packet switching in radio channels: Part I – carrier sense multiple-access modes and their throughput delay characteristics.* IEEE Trans. Commun. vol. COM-23, no. 12, 1975

[9] TOBAGI F. A., KLEINROCK L., *Packet switching in radio channels: Part II – the hidden terminal problem in carrier sense multiple-access modes and the busy-tone solution.* IEEE Trans. Commun. vol. COM-23, no. 12, 1975

[10] WEINMILLER J., WOESNER H., EBERT J-P.., WOLISZ A., *Analysing the RTS/CTS Mechanism in the DFWMAC Media Access Protocol for Wireless LANs*, IFIP TC6 Workshop Personal Wireless Communications, April 1995, Prague Czech Republic

[11] WOŹNIAK J., NOWICKI K., *How the Usage of RTS/CTS and HNA Mechanisms Can Improve the Fairness of Operation of IEEE 802.11 and HIPERLAN Networks In the Presence of Hidden Stations?*- 6th Polish Teletraffic Symposium 22-23 April 1999 Szklarska Poręba

[12] WU C., LI V. O. K., *Receiver-initiated busy-tone multiple access in packet radio networks.* ACM SIGCOMM 87 Workshop: Frontiers in Computer Communications Technology, Stowe, VT, USA 11-13 Aug. 1987

An Overview of Activities on Wireless Networks in the European Project COST 257

Wojciech Burakowski (Poland), Udo Krieger (Germany), Kenij Leibnitz (Germany), Andrzej Beben (Poland), Michela Meo (Italy), Tolga Ors (United Kingdom), Jorge Garcia-Vidal (Spain), Markus Fiedler (Sweden)

COST 257 Project participants

E-mail: wojtek@tele.pw.edu.pl, leibnitz@informatik.uni-wuerzburg.de

Key words: wireless networks, modelling, performance, network planning, analysis

Abstract: The paper summarises the work on wireless networks inside the COST 257 project, entitled "*Impact of new services on the architecture and performance of broadband networks*" and chaired by Prof. Tran-Gia of the Technical University of Würzburg (Germany). This project was established for 4 years (1996-2000) and collected the researchers from 16 European countries represented by 32 organisations. Its topics are modelling, performance, network planning and analysis of present and future wireless systems.

1. INTRODUCTION

Extending network-based services to wireless mobile subscriber access was an important milestone in the global telecommunication network evolution. Spectacular success of the cellular telephone networks (e.g. GSM – Global System for Mobile) has proven that all, today's and future, network technologies should offer wireless and mobile capabilities.

Enhancement of services supported by the wireless networks is planned to follow evolutionary scenario comprising four main phases, 1. However, introduction of new services is conditioned by available bandwidth of radio access channel. Today's digital cellular mobile network belongs to the 2nd generation (2G) of wireless networks offering very low to low bit rate communication services for voice and data. Further network evolution is on the way towards including packet data services (G2+), Narrow Band Integrated Services Data Network (N-ISDN) capabilities (G3), until, from today's perspective, Broadband ISDN (B-ISDN) services are offered (G4).

A general wireless network architecture assumes three main levels, which are the access network, the core transport network and the service network 2. In the access network, the base station (BTS) provides radio connectivity to the mobile stations (MS). For this purpose, a specialized radio access protocol is implemented in both the base and mobile station. Recognized solutions for such protocols are based on TDMA or CDMA schemes. In the TDMA, for a single MS station a number of time slots in the frame are assigned for the transmission while in the CDMA, all MS stations could generate traffic at the same time using different orthogonal codes 3. Optionally, the access network can contain concentrators (C) to connect a number of BTS stations to the core network. One can distinguish between two types of access networks, indoor and outdoor. In the indoor network the distance between MS and BTS stations is relatively small and typically limited up to 100 meters. On the contrary, the outdoor network covers rather large area, up to few kilometers. The size of this area strongly depends on such parameters as number of MS stations and signal propagation conditions.

The core network is a wide area network that is designated for transporting user and signaling information, and is typically built on cable links (although some wireless links can also be applied, like satellite or radio relay). This network could support a variety of communication services, depending of the involved technology. In general, the access and the core networks are not necessary to be build using the same telecommunication technology. However, for supporting end-to-end network services with adequate quality of service one technology is strongly desirable. For instance, the complete set of broadband services require ATM in both the access and the core network.

The service network contains highly specialized servers that support mobility management and provides application specific processing, e.g. for speech transcoding, voice mail, message and facsimile handling.

In the wireless networks the QoS provisioning problem is more challenging than in the fixed network. This is due to the worst wireless channel characteristics, mainly caused by limited bandwidth and relatively high (and temporary variable) bit error rate. In order to cope with the transmission errors some protection mechanisms are required at the physical or data link layer. However, the implementation of such mechanisms produces additional overheads decreasing effective radio link capacity. As a consequence, availability of resources the network dedicates for a given connection could be variable. Additionally, in the access network a number of connections share the same radio link, and this requires application of special MAC protocols. The problem is becoming more complicated when the network additionally supports user mobility. When a customer requires handover, the network should also change the resource allocation to new BTS station.

2. NARROWBAND NETWORKS

Present digital cellular systems are considered to be part of the second generation of wireless networks (G2). While these systems were primarily designed to offer voice services, they provide only limited capabilities for data transmission at low and medium bit rates. Most prominent examples for narrowband networks are the Global System for Mobile communications (GSM) in Europe as well as IS-54, IS-95, and Personal Digital Cellular (PDC) in North America and Japan. Additionally, some of the second generation systems like Digital European Cordless Telecommunications (DECT) and Personal Handyphone Systems (PHS) have been introduced to offer wireless services in residential and office environments.

All second generation wireless systems support circuit switched voice and data services with basic rates typically ranging around 9.6 kbps. With the growing needs for higher rates, new services are currently being developed that permit multiple allocation to the physical resource. Extensions to GSM like the High Speed Circuit Switched Data (HSCSD) or GPRS (General Packet Radio Service) are considered to be in a transition phase (G2+) towards third generation systems (G3) allowing bit rates up to 170 kbps. In addition, enhancements to GSM itself like the introduction of a new modulation scheme under the name of EDGE (Enhanced Data Rates for GSM Evolution) further indicate the evolution of narrowband systems.

2.1. Planning of Narrowband Wireless Networks.

Due to today's tremendous customer demand and rapid growths of mobile networks, the need for a systematic planning methodology has become essential. Furthermore, knowledge about customer behavior and measured traffic data allow for detailed planning procedures. In conventional cellular planning, the planning process is driven by radio coverage considerations, i.e., selection of cell site locations, frequency planning, antenna design, etc.

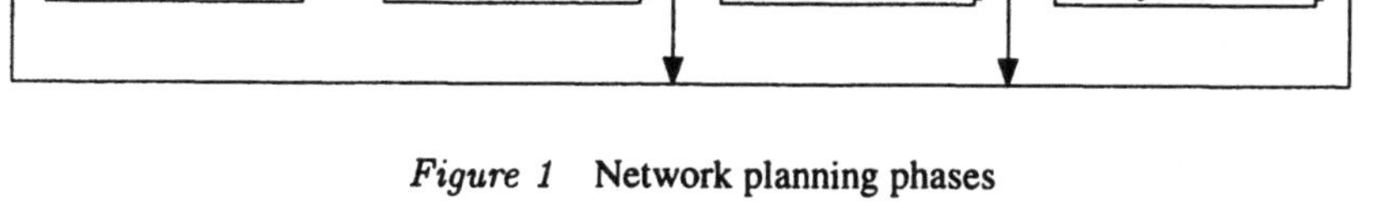

Figure 1 Network planning phases

Basically, the algorithm used for cellular planning consists of four phases, cf. Fig. 1.

1 In the *Radio Network Definition* phase, an experienced radio planning engineer chooses cell site locations based on his knowledge and planning experience.

2 Then, the *Propagation Analysis* evaluates the radio coverage of the area using field strength prediction methods.

3 If the coverage requirements are met, the expected number of traffic channels is calculated and a *Frequency Allocation* is performed.

4 If the frequency plan can be computed, the network performance is evaluated in the *Radio Network Analysis* phase by computing some quality of service (QoS) measure in the cells.

From the above description it is apparent that the consideration of the customer traffic influences the network planning approach only to a very limited extent. Thus, a new approach to performing a truly *demand oriented* network planning has been introduced that takes as input a characterization of customer demand and incoming traffic by discrete points, the *demand nodes*.

The core component of the new integrated design concept are the automatic network design algorithm *Set Cover Base Station Algorithm* (SCBPA) and a traffic characterization procedure which generates a set of demand nodes. The first step in this approach is to create a distribution of demand nodes based on estimations of the demand in a certain coverage area. Then, the SCBPA uses a greedy heuristic which selects the optimal set of base stations that maximizes the proportion of covered traffic, i.e., the number of demand nodes which measure a field strength level above a certain threshold value.

One of the key components of the system model for network performance analysis is a model of the demand node pattern. The other major issue is to characterize the behavior of the users at call level and obtain a measure of the performance of the system in terms of subjective *Quality of Service* (QoS) of the user. Both items will be described in greater detail in the following sections.

2.2. Description of the Spatial User Demand.

In this section, we will describe some methods on characterizing the spatial user distribution. At first, we will describe in more detail the notion of a demand node and give a definition as well as a method for generating a demand node pattern from geographical and demographical input data. When such data is unavailable, it is possible to use the realization of a stochastic spatial process. This theoretical approach will also be covered later in this section.

The Demand Node Concept. A *demand node* represents the center of an area that contains a quantum of demand from teletraffic viewpoint, accounted in a fixed number of call requests per unit, see 4, 5. Such demand nodes represent the user demand discretized both in space and demand.

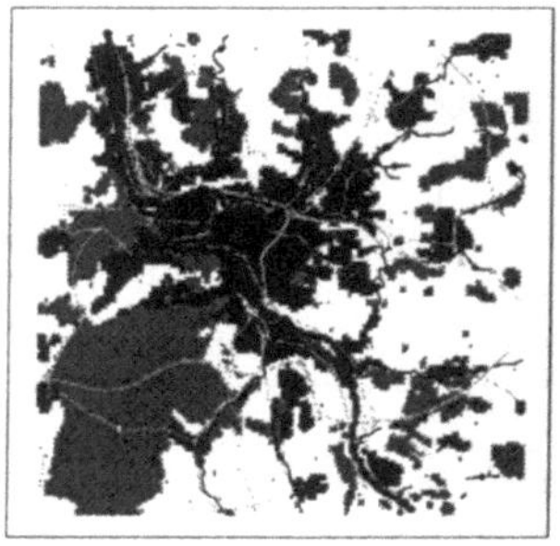

Figure 2 Geographical and demographical data.

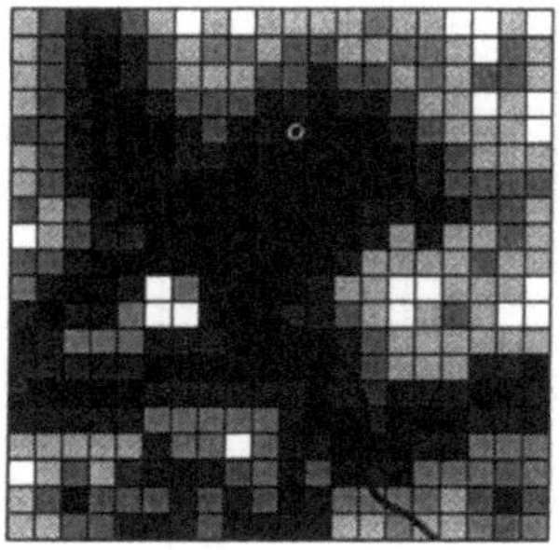

Figure 3 Traffic matrix.

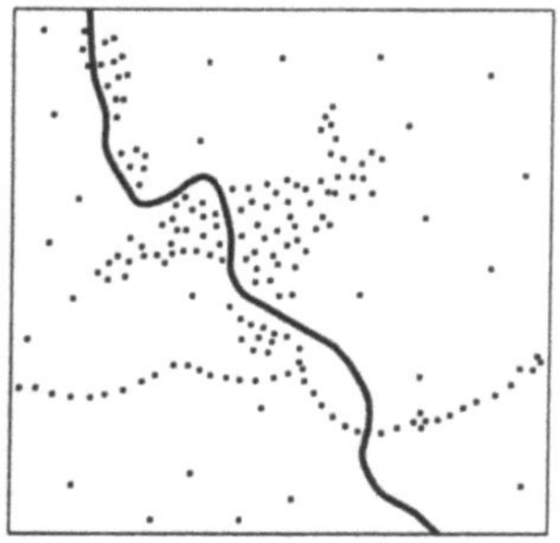

Figure 4 Demand node distribution.

Empirical Distributions. The sequence of generating a discrete traffic estimation is shown in Fig. 2, 3, and 4. Fig. 2 depicts the input of the estimation algorithm. It contains land usage data that is usually available for the specific area. The colors represent different categories of land usage, e.g. urban, suburban, forest, water, or open areas. Each class is assumed to generate a certain amount of traffic and the traffic matrix is obtained by superimposing the traffic from the different land usage classes active on one area element (Fig. 3). The darker an element of the traffic matrix is, the higher the anticipated demand in the corresponding area element. Then, the matrix representation is being transformed into a spatial discrete point pattern by using a clustering algorithm. A partitioning method has been described in 5. The result of this partitioning is shown in Fig. 4.

Spatial Poisson Process. While the empirical demand node patterns are the most desirable for network planning, often the data is not available to create the traffic matrices. In this case, theoretical stochastic processes are used to describe the user distribution. Such a process is in general a random variable, which takes random choices of mappings from a Borel set to a counting measure, the number of simple points.

The simplest distribution of demand nodes is described by a spatial Poisson point process. In this case the number of points in any Borel set on the plane follows a Poisson distribution, depending on the area of the set and the intensity of the process. If the intensity measure of the process is independent of the location, it is called *homogeneous*. Such a process is depicted in Fig. 5.

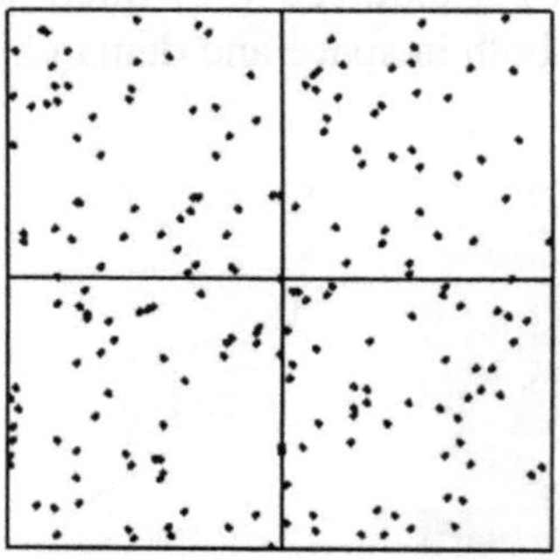

Figure 5 Spatial Poisson process.

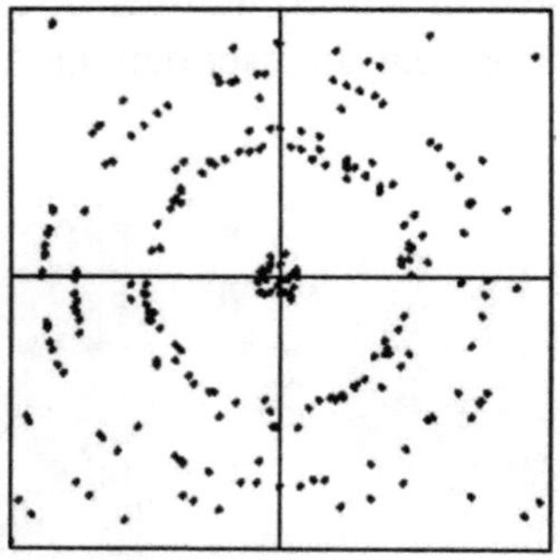

Figure 6 IPhP³ process with hyper-exponential distribution.

IPhP³ process. The *Isotropic Phase Planar Point Process* 6,7 allows for characterization of a large number of spatial distributions and can be considered a general case of spatial process. Due to its construction, this process can introduce cluster effects and dependence between sets of points. A realization of the process with hyper-exponential distribution of the inter-point distance is given in Fig. 6. Here, clearly a cluster effect in circular shape can be observed.

2.3. Modeling of Subjective QoS. In this section, we present a basic model aiming at answering the question: How does the cluster structure influence the subjective QoS of a customer in a mobile network? If the network planning does not take into account any customer clustering, the subjective QoS in terms of *call blocking probability* experienced by a specific test customer will decrease in areas with customer concentration.

For this reason, the cell is described by a finite source model in which the number of sources is obtained from the spatial locations of the customers. In the simple case 4, if the customer is blocked, he will remain idle. This model is extended in 8 to include the repeated attempt behavior of the customer. The reaction of the user when experiencing call blocking can influence his QoS significantly.

Customer Traffic Process. Fig. 7 shows the model of the cell where the customers can either be in idle or in active states. After call termination or rejection, the user will remain idle until generating the next call. The model is therefore a standard loss system with K servers (number of channels) and X sources (random number of customers). The novelty of this approach is to derive the distribution of X from the spatial traffic description. In 4, it is

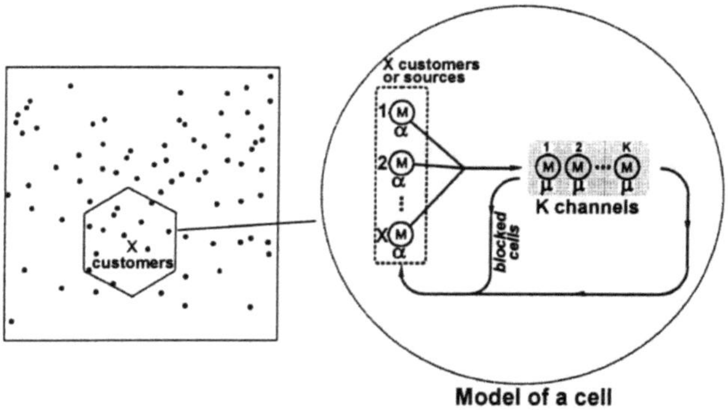

Figure 7 Single cell with finite number of customers

shown that there is a degradation of the QoS when the customer population is clustered and the planning did not take such structure into account.

User Retrial Phenomenon. The model can be extended to include a phenomenon quite common in daily life: A user whose call is blocked will immediately retry calling instead of giving up. This causes the network load and thus the blocking probability to rise even further — a snowballing effect of call arrival processes can occur, which leads to dramatic degradations of the call completion performance of single switching systems, and subsequently of the whole network. Fig. 8 shows the modifications of the model from Fig. 7, where Θ denotes the *retrial probability*. The analysis is carried out by solving two-dimensional Markov chains in a recursive way (8).

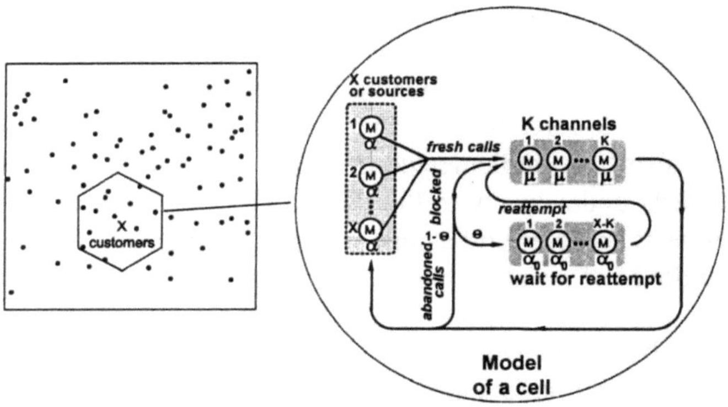

Figure 8 Single cell with finite number of customers and modeling of the retrial phenomenon.

Fig. 9 illustrates the impact of the retrial probability on the blocking probability of first attempts for different ratios of mean call holding time and mean wait-for-reattempt time α_0/μ. It can be clearly seen that the blocking probability of the first attempt is higher than without repeated attempt. Comparing to the case without retrials ($\Theta = 0$), it is clear that we cannot neglect the repeated attempt phenomenon.

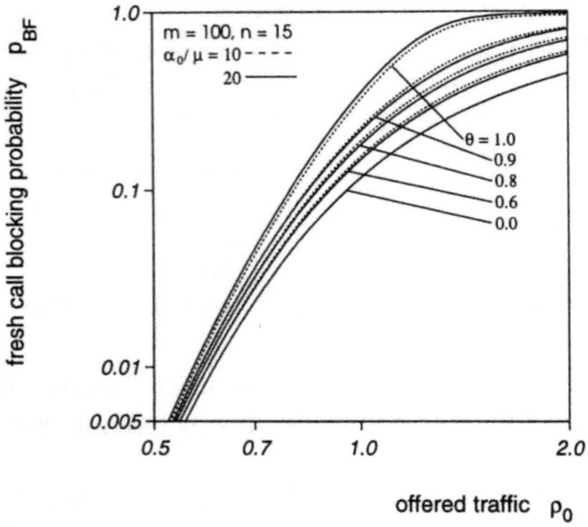

Figure 9 Single cell with finite number of customers.

This phenomenon also holds in a model in which hand-over calls are prioritized over fresh calls by the use of guard channels.

Integrated Services Call Level Model. A service request from a mobile user to a base station may be due either to the generation of a new call or to a handover request. Since handover failures force the termination of a call in progress, these events are considered to be worse than new calls blocking, whose effect is just to force the user to repeat his access request at a later time.

Two mechanisms are implemented in order to favor handover requests over new call requests under heavy traffic conditions ?. First, priority is given to handover requests by reserving a small number of channels for them; second, when the handover of a multimedia call is requested, the base station tries to accommodate the entire call, but, if this is not possible, the two components of the call can be decoupled: the voice connection is accepted, if possible, while the data connection is temporarily suspended, waiting to be resumed as soon as enough channels are available in the cell to serve the entire multimedia call. Decoupling a multimedia call because of a partially served handover can improve the quality of service, mainly because it allows the user to continue at least partly his communication. To facilitate recombination of decoupled multimedia calls, new calls are not accepted as long as some active decoupled connections exist in the cell.

The cell state, in any instant, is determined by the number of currently active connections for each class of traffic. A continuous-time Markov chain is derived where transition rates are determined from the rates of the arrival processes of new call and handover requests, and from those of the processes of call completion and outgoing handover requests.

As a basic scenario for validation, a configuration with two classes of service was used, comprising voice calls and data calls which require the allocation of 4 channels. Curves showing the average carried traffic for the two classes of service, the average number of busy channels and the average blocking probability for new call and handover requests, were produced using both the analytical model and a simulator. The comparison of the curves clearly indicates that the approximations introduced in the model development do not alter the numerical results significantly.

Other results were also obtained with different configurations of the cell under investigation. Scenarios with voice, data and multimedia connections were considered, with different values of the number of channels required for each data call; the performance of a system in which some channels are reserved to data connections was evaluated, assessing the performance improvements for systems in which the decoupling of multimedia connections is introduced.

2.4. Dimensioning of Voice Traffic Links. Another important aspect in the planning of cellular networks is the appropriate dimensioning of the links transporting the voice traffic. This is especially important when migrating towards future mobile network generations where also data, video, and multimedia services will be supported. In such systems the CDMA access method with variable bit rate vocoders will be employed, resulting in the problem of efficiently transporting streams of low and variable bit rate over a high speed transport network, usually operating with ATM technology.

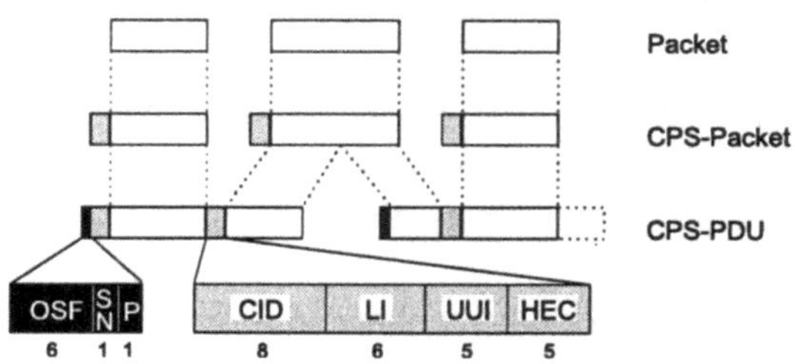

Figure 10 Packetization of voice packets with AAL2

The application of ATM Adaptation Layer Type 2 (AAL2) is intended, supporting the multiplexing of several narrowband connections on one ATM connection, see Fig. 10. After encapsulating the voice packet in a CPS PDU, several such CPS packets are fitted into a single ATM cell. In case that the remaining space in the ATM cell is not sufficient to include a CPS packet, its payload is split up in two cells. Additionally, a timer serves to keep the packetization time within a certain delay requirement.

In 9 and 10 studies were performed that investigated the suitability of AAL2 for narrowband CDMA speech connections. Here, the packetization and traffic shaping was modeled as discrete time Markov chain. The transitions were given as recursive state transition equations. From the analysis the distribution of the delay of the speech packet was derived. The results for an E1 link showed that AAL2 is capable of transporting about 80% of the traffic compared to the traffic without the overhead of AAL2. The reason for this loss in capacity is attributed to the additional CPS packet headers. This example illustrates the need for detalied modeling of the transport network in order to achive a correct dimensioning of the system.

3. BROADBAND WATM NETWORKS

In 1996, the ATM Forum Wireless Working Group started activities to develop standards for the wireless ATM network (WATM). Recall that ATM was originally designed for fibre-optic transmission links characterised by extremely low error rate (10^{-12} or less). Therefore, adaptation of ATM into the wireless requires updating existing standards by adding radio access layer as well as user/terminal mobility management. These new standards should allow the WATM network to provide the same set of communication services as in the wired network (with similar QOS requirements).

The reference configuration for the WATM network with all WATM application scenarios 11, 12, shows Fig. 11. These configurations differ in terms of terminal mobility, wired or wireless access, mobile or fixed switches, internal ability for establishing ad-hoc networks and co-operation level with PCS (*Personal Communication Systems*) systems.

The configurations # 1 and # 2 refer to two distinct components of WATM: the radio access technology, and enhancement of existing ATM technology to support end-system mobility, respectively. The rest of recommended configurations correspond to the enhancement WATM for supporting mobile switches, building ad-hoc networks, supporting PCS access and internetworking PCS-ATM.

The WATM network demands an enhanced protocol stack comparing to cable ATM network. The required protocol stack, which should be implemented in the terminals, access points and/or ATM switches, depends on the assumed

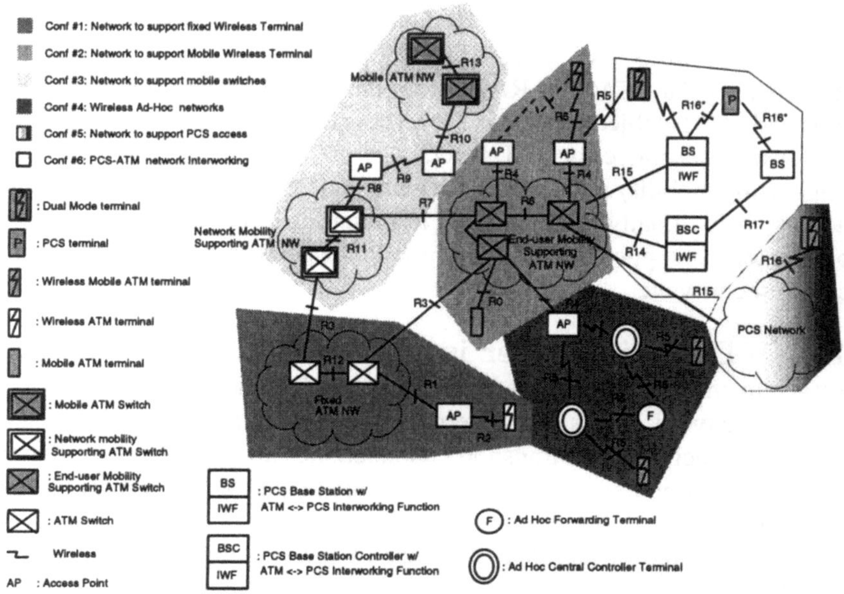

Figure 11 Reference configuration for WATM

network configuration. These enhancements are related with wireless access and mobility management. Wireless access provisioning requires some modifications on the physical layer while supporting mobility demands new functions implemented at the ATM layer and in the signaling protocol.

The reference protocol stack to cope with wireless and mobility is depicted on Fig. 12 12.

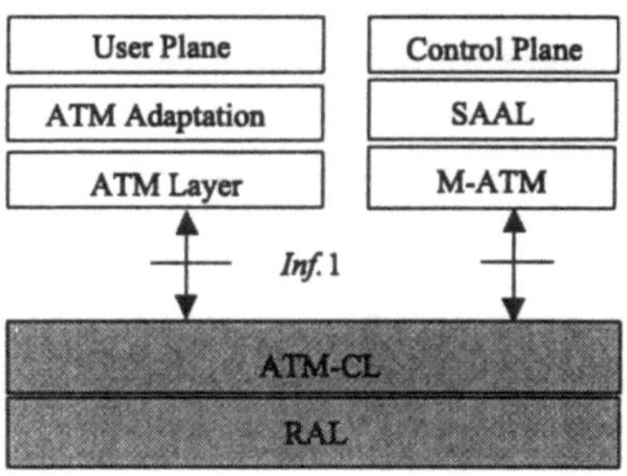

Figure 12 Reference protocol stack

The objective of the RAL (Radio Access Layer) layer is to support data transport over radio links with satisfactory quality. It includes the wireless

MAC, LLC and PHY layers. The RAL may be generic, designed to support multiple higher protocols such as ATM, IP, etc.

Between regular ATM and RAL layers the ATM CL (Convergence Layer) layer is added. The objective of this sublayer is to provide ATM cell conversion to the format acceptable for RAL layer.

Supporting mobility management demands some enhancements in signaling protocols. Additionally, in the mobile terminal local M-ATM layer should be implemented for supporting different mobility aspects, e.g. handover, terminal tracking, security etc.

3.1. Data link layer issues. Considering the transport of messages in an advanced packet-switched network at the data link, network interface and network layers over an error-prone communication channel, e.g. a wireless ATM network, the transport characteristics are additionally influenced by handover procedures according to the implemented terminal mobility management 13.

Furthermore, the hierarchical structure of the existing protocol stack, the segmentation of messages and the correlation of errors at the wireless physical layer as well as the error recovery at the wireless LLC layer have a strong impact on the transport performance.

The network interface layer, called block layer here, coincides with the ATM adaptation layer and provides a service-specific bearer service over an error-prone communication channel for the network and transport layers on top of it, for example LLC/SNAP encapsulated IP frames (cf. 14, Sec. 4.1.2, p 202f). An Automatic-repeat-request (ARQ) protocol such as Selective Repeat (SR) or Go-back-N (GBN) and their ramifications can be applied both at the wireless data link layer as well as at the service-specific convergence sublayer (SSCS) or at the common part convergence sublayer (CPCS) of the ATM adaptation layer (AAL) to cope with the defective transport of ATM cells and to guarantee a secure delivery of messages to the higher layers cf. 15, 16, 17.

3.1.1 Objectives. Regarding the efficiency of the data transport a large block length n is required while error considerations demand short blocks. Therefore, we suppose that the size of the block PDUs can be varied to some extent, that the transport channel of cell PDUs is error prone due to noise, failures of the transmission equipment, data loss, as well as customer and terminal mobility and that a Go-back-N (GBN) ARQ protocol with window size m is applied at the block layer to guarantee a correct delivery of the block PDUs free from losses, cf. 17, p. 129f.

The question arises how the performance of a GBN protocol subject to the segmentation of messages and the correlated errors of the PDUs, particularly the additional load due to retransmissions, can be determined by analytic means and how the corresponding parameters should be selected. Then it may be possible to perform a self-tuning of the block length and window size processes under the control of the error detection process at the cell level.

3.1.2 Analysis of the Go-Back-N performance.

To analyze the impact of correlated errors at the block and cell levels on the performance of the GBN ARQ protocol running at the block level we assume in accordance with observations in real networks, that the transport errors of consecutive blocks b_i, b_{i+1}, $i \geq 1$, are correlated (cf. 15, 16, 18, 19). We assume that a cell at position $j, 1 \leq j \leq n$, in b_{i+1} is damaged with probability s if its counterpart was already damaged at position j in b_i. With probability p it is transported without an error if the cell at the same position in the previous block was received without an error (see Fig. 13).

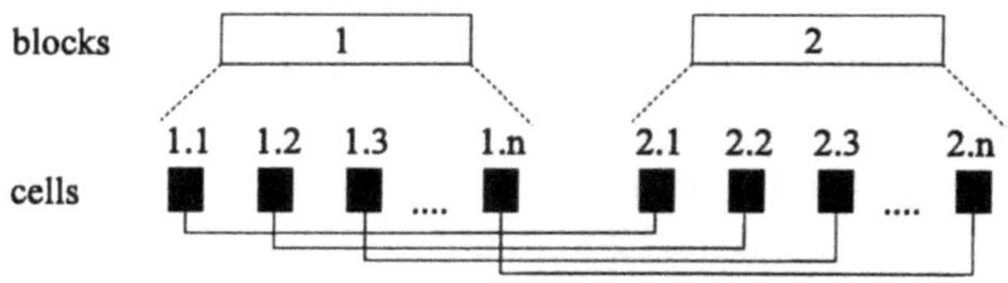

Figure 13 Correlation of errors during the cell transport

The corresponding error process at the cell level is modeled by a binary discrete-time Markov chain (DTMC) $\{X_j(t), t \geq 0\}$ on a probability space $(\Omega, \mathcal{K}, \mathbb{P})$ with state space $\Sigma = \{0,1\}$ and the states $X_j(t)(\omega) = 0, \omega \in \Omega$, for error-free and $X_j(t)(\omega) = 1$ for defective cell transport of the t-th block. $X_j(0)$ represents some initial condition of the process. The DTMC is described by a transition probability matrix (t.p.m.)

$$P_C = \begin{pmatrix} p & 1-p \\ 1-s & s \end{pmatrix} = \begin{pmatrix} p & q \\ r & s \end{pmatrix}$$

with $q = 1 - p, r = 1 - s, 0 < p < 1, 0 < s < 1$ (see Fig. 14).

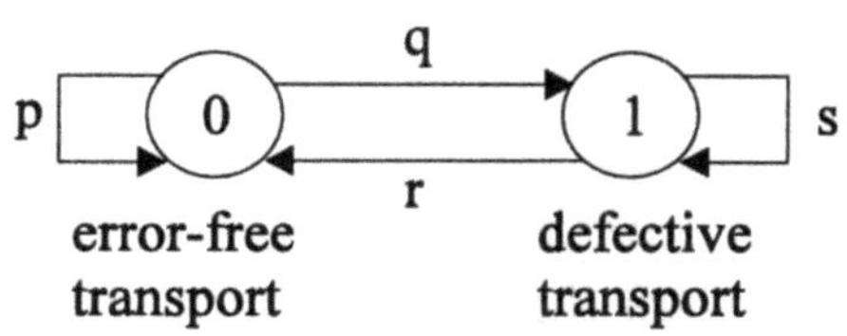

Figure 14 Markov model of a single cell error process

However, a more complicated correlation structure depending on the block length n by some function $\chi(n) \in \mathbb{N}$ can be incorporated into a similarly structured two-state model using the t.p.m.

$$P_C(n) = \begin{pmatrix} p(n) & q(n) \\ r(n) & s(n) \end{pmatrix} = \begin{pmatrix} \hat{p} & \hat{q} \\ \hat{r} & \hat{s} \end{pmatrix}^{\chi(n)}$$

with $\hat{q} = 1 - \hat{p}, \hat{r} = 1 - \hat{s}, 0 < \hat{p} < 1, 0 < \hat{s} < 1$, e.g. $\chi(n) = n + 1$ yields

$$\begin{aligned} P_C(n) &= \begin{pmatrix} \hat{p} & \hat{q} \\ \hat{r} & \hat{s} \end{pmatrix}^{n+1} \\ &= \frac{1}{\hat{r} + \hat{q}} \left[\begin{pmatrix} \hat{r} & \hat{q} \\ \hat{r} & \hat{q} \end{pmatrix} \right. \\ &\qquad \left. +(1 - \hat{r} - \hat{q})^{n+1} \begin{pmatrix} \hat{q} & -\hat{q} \\ -\hat{r} & \hat{r} \end{pmatrix} \right] \end{aligned}$$

and reflects a nonlinear dependence of the correlation structure on the block length (cf. 15).

In the following we illustrate the proposed approach using only the simplest model. To simplify the analysis and to proceed to a tractable model, the individual error processes X_j, $1 \leq j \leq n$, at each cell position j in consecutive blocks are assumed to be independent of each other and in steady state. Applying an aggregation-disaggregation argument, its behavior is approximated by a t.p.m. of the form

$$\Gamma(n) = \begin{pmatrix} p^n & 1 - p^n \\ \omega(n,r,p) & 1 - \omega(n,r,p) \end{pmatrix} \in \mathbb{R}^{2\times 2} \tag{1}$$

with

$$\omega(n,r,p) = \frac{1}{2^n - 1} \cdot [(r+p)^n - p^n] > 0 \tag{2}$$

where the range of $s, p \in (0,1)$ has to be limited such that $0 < \omega(n, 1 - s, p) < 1$. The steady-state block error probability of the aggregated DTMC $Z(t)$ is given by:

$$\begin{aligned} \epsilon &= \epsilon(n,r,p) = \mathbf{P}\{Z = 1\} \\ &= \frac{1 - p^n}{1 - p^n + \omega(n,r,p)} \in (0,1) \end{aligned}$$

We assume that the return channel sending the positive and negative acknowledgements of the ARQ protocol is not disturbed, that the transport channel for block PDUs is working in a slotted manner where a PDU transmission time constitutes one time slot, and that for a window size m the round-trip propagation delay is constant and equal to an integral number $m-1$ of time slots, cf. 15. Then the mean delay

$$D = \frac{m}{\eta}$$

of the block transport (cf. 15, (2.10), p. 232, 16, (30), p. 725). and the throughput efficiency η of the Go-Back-N protocol with window size m for the aggregated correlated error process $\{Z(t), t \geq 0\}$ in steady state with the t.p.m. $\Gamma(n)$ are determined by:

$$\eta(n,s,p) = \left[1 + \frac{m(\omega(n,1-s,p)+1-p^n)\cdot(1-p^n)}{\omega(n,1-s,p)\left[1-(p^n-\omega(n,1-s,p))^m\right]}\right]^{-1}$$

$\eta(n,s,p)$ is a function of the window size m, the block length in terms of the number n of consecutive cells and the transition probabilities p between error-free cell transports and that one s between defective transports, respectively.

Some results are shown in Fig. 15 and 16. Here we assume that the probability $(P_C)_{10} = 1-s = r$ of a transition from an unsuccessful to a successful cell transport between consecutive blocks is given by $r = s = 0.5$ and the probability $(P_C)_{01} = 1-p = q$ of a transition from a successful to an unsuccessful cell transport between consecutive blocks as well as the number n of cells per block are varied.

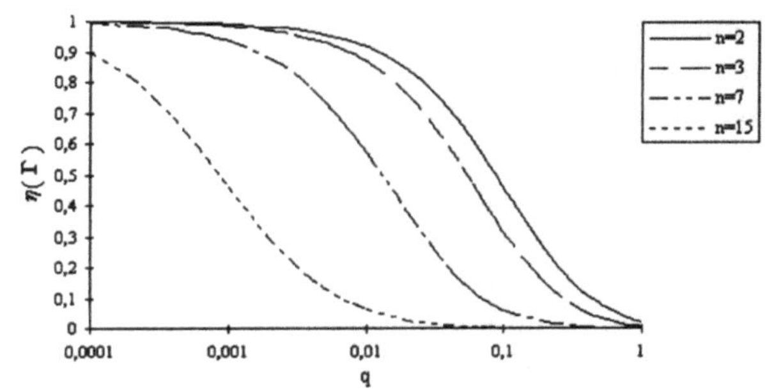

Figure 15 Throughput efficiency of GBN ARQ for varying transition probability q and window size $m = 4$

The sketched model can be used as first simple approach to study the impact of varying block length, segmentation and correlated errors on the performance of an ARQ protocol.

3.2. Satellite MAC protocols.

In this section we analyze the performance of an Adaptive Random-Reservation Medium Access

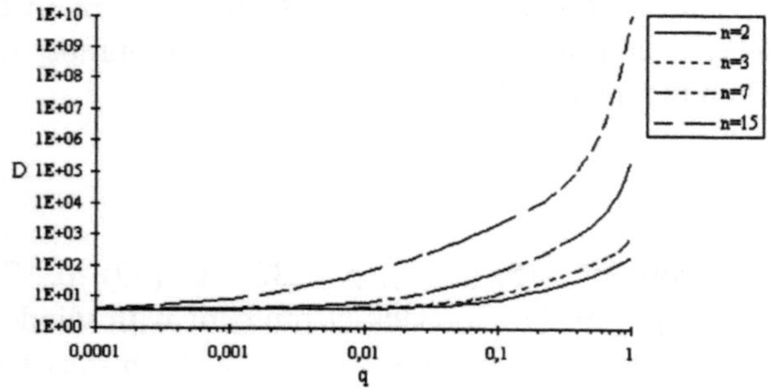

Figure 16 Block delay of GBN ARQ for varying transition probability q, $r = 0.5$, message length $n \in \{2, 3, 7, 15\}$ and window size $m = 4$

Control (MAC) protocol which can support all ATM service classes while providing the required Quality of Service (QoS).

Our study focuses on parameter optimisation of the multiple access schemes for ATM over a GEO satellite with on-board processing capabilities, considering various traffic mixes of CBR *(Constant Bit Rate)*, rt-VBR *(real time Variable Bit rate)*, nrt-VBR *(non-real time Variable Bit Rate)*and UBR *(Unspecified Bit Rate)*.

It is shown that maximum throughput can be achieved by using this access scheme. A TDMA access protocol combining both Random Access and Demand Assignment Multiple Access (DAMA) is particularly suited for a scenario with a high number of terminals with very bursty UBR traffic (e.g. web browsing). UBR sources with short burst length access the slots remaining after the reservation procedure by random access which drastically reduces the slot access delay, at the expense of lower utilisation. However for UBR sources with burst sizes consisting of several ATM cells, reservation access provides higher throughput but the access delay is considerable longer.

The adaptive MAC protocol was designed to allow statistical multiplexing of ATM traffic over the air interface, especially for the independent and spatially distributed terminals. It is shown that the potential user population which can be served is considerably increased by statistically multiplexing bursty traffic over the air interface.

3.2.1 Framework for proposed MAC protocol.

Design objectives. The MAC protocol has to be designed to allow statistical multiplexing of ATM traffic over the air interface, especially in the uplink for the independent and spatially distributed terminals. The following design objectives are taken into consideration:

- maximize the slot utilization, especially for bursty traffic
- guarantee the QoS requirements for all service classes
- maximize frame efficiency by minimizing overheads

The minimization of overheads is not an easy task, especially for ATM which was designed for channels with very good error characteristics (Bit Error Rates around 10^{-10}). To minimise cell loss over the satellite link, channel coding has to be used to make the transmission more robust. A LLC header to facilitate error recovery mechanisms is optional and not in scope of this study. Finally a satellite specific header with satellite routing and wireless resource management fields is added to form a MAC packet as shown in Fig. 17.

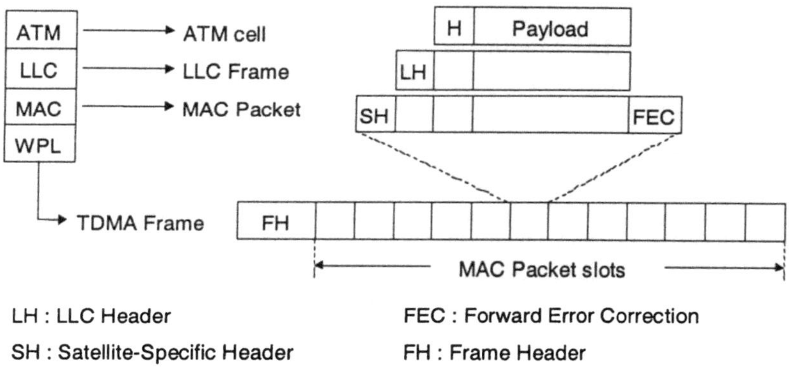

Figure 17 Encapsulation of ATM Cells to MAC Packets and mapping to TDMA Frame

Access schemes. MAC layer access schemes can be typically categorised into four classes: Fixed Access, Random Access, Demand Assignment Multiple Access (DAMA) and Adaptive Access. The first three techniques have evolved to meet the needs of constant high traffic with long duration's, sporadic traffic with short to medium duration's, and sporadic traffic with long duration's, respectively. Finally adaptive access is used to meet the needs of multiple media which consists of traffic with various characteristics. Thus to meet the design objectives an Adaptive Access mechanism seems to be the best choice.

Mapping of ATM service classes onto MAC service classes.

To simplify the conceptual design of the MAC protocol, ATM service classes 20 can be mapped onto MAC service classes.

Mapping of CBR service category. Fixed-Rate DAMA is ideal for connections with a constant bit rate such as the CBR service class in ATM networks. Before a connection is set-up, the terminal and satellite negotiate the Quality of Service (QoS) parameters. These QoS parameters determine the characteristics of the connection. Since the parameters will not be modified during the connection, the amount of bandwidth allocated for that connection will not be changed until the connection is terminated. For ATM CBR connections the Peak Cell Rate is allocated to the terminal.

Mapping of rt-VBR service category. Real-time Variable Bit Rate services can also be supported with fixed-rate DAMA. For real-time services, the amount of bandwidth assigned to the connection should be close or equal to the Peak Cell Rate (PCR) to avoid cell delay. The major drawback of this scheme is that a major portion of the bandwidth is wasted when the cell transfer rate is lower than the assigned bandwidth. The major difficulty to employ variable-rate DAMA in ATM satellite systems is the effect of the large propagation delay. The computing and negotiation process between the satellite and the terminal may be too long for rt-VBR services and result in unacceptable QoS. The use of variable-rate DAMA for rt-VBR is only possible if the arriving traffic can be predicted one hop delay in advance. Since this is not possible except in some special cases fixed-rate DAMA will be used for rt-VBR. A scenario where fixed-rate DAMA is efficient for rt-VBR services is when the terminal can multiplex traffic from multiple services. In this case the aggregate traffic can be approximated as a constant cell flow by using a small amount of shaping.

Mapping of nrt-VBR service category. VBR services which are not time sensitive can be assigned an effective bandwidth which is between the mean cell rate and PCR. Since the required bandwidth of VBR sources changes with time, there may be instants when the cell transfer rate is higher than the amount of bandwidth (effective bandwidth) assigned to that connection. In this case cells can be buffered in the terminal and in case the queue exceeds a certain threshold more bandwidth can be requested. Thus using variable-rate DAMA the bandwidth of a connection can be adjusted according to the change of the data transfer rate.

Mapping of UBR service category. No numerical commitments are made for the UBR service class and this service category is intended for non-real time applications. UBR services could be supported by variable-rate DAMA. However the fact that this service class has the lowest priority (because no commitments to CLR are made) has to be considered. We propose that UBR could transmit data directly to the unoccupied data slots without reservation. The unreserved slots are broadcasted on the downlink to be accessed by random access. This is particularly appealing for bursty interactive services with short duration, for which the long slot reservation delay is unacceptable.

3.2.2 The random-reservation adaptative assignment protocol. The TDMA frame of the adaptive assignment protocol is divided into Reservation slots, Control slots, Data slots and Random Access slots, as shown in Fig. 18. The protocol is based on the proposals by 21, 22, 23 with modifications to achieve the design objectives for multi-service networks.

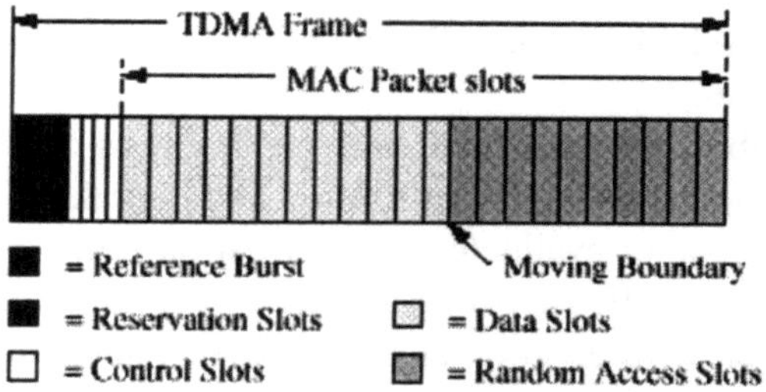

Figure 18 Frame Structure

A Reservation slot is that period of time in which terminals report their requests to the reservation unit. There are only a few Reservation slots available and a terminal selects one at random without knowing whether another station is using the same slot. If more than one terminal selects the same reservation slot, a collision occurs and terminals have to retransmit after waiting for a mean retransmit waiting time determined by the collision resolution algorithm. The MAC protocol ensures that the collision probability stays low. The reason for using reservation slots is because ATM networks support different services which have different loss and delay requirements.

If a single request was received for a reservation slot (successful request), the on-board wireless resource management module tries to allocate the necessary Data slots. If no Data slots are available, the request can either be blocked (called blocking probability) or queued. We propose to queue successful requests in a prioritised queue so that the terminal does not need to compete with other terminals for a reservation slot again. By queuing successful reservations, requests can be allocated data slots according to their priorities.

Once the Data slots are reserved (successful reservation), an acknowledgement is transmitted to the terminal in TDM mode on the downlink frame. Data slots represent the part of the frame in which a terminal can transmit its message after a successful reservation. In every frame there are many data slots and the on-board wireless resource management module will assign data slots to a particular successful request. A data slot is assigned to at most one terminal and therefore there is no possibility of collision.

On the other hand Random Access (RA) slots represent the part of the frame in which terminals can transmit without the need of making a reservation. The slots available for random access are broadcasted in the downlink frame. This part is for services which don't want to wait for the lengthy reservation procedure. In random access mode it is not possible to guarantee a certain QoS to users although the protocol will try to minimise the number of collisions to maximise throughput by using an adaptive collision resolution algorithm. Random Access should only be used by UBR sources with relatively small burst length since RA terminals are not allowed to reserve slots and have to content for each MAC packet.

Unless the number of reservation slots per frame is carefully adjusted the result would be either low capacity utilisation and long delays (too many reservation slots, less capacity available for information transmission) or network backlog (too few reservation slots resulting in successive collisions and high delay). The number of reservation slots should be fixed for system behaviour where the number of collisions can be controlled by broadcasting a message in the downlink that services with lower priority should not send/resend requests till the collisions have been resolved. Our analysis has shown that two reservation slots provide adequate performance. However when the number of collisions can't be controlled new reservation slots can be added by reducing the number of control slots.

The requests for dynamic slot allocation are done using the Control Slots which are assigned, on a round-robin basis to all terminals which request the variable-rate DAMA MAC class. The number of control slots is set to eight to minimise the frame overhead.

The satellite frame introduces a constant delay equal to the frame length, on the cells of a stream connection. Therefore the selection of the frame size should be small enough to satisfy the delay limit of real-time services (400ms) 24 taking into account the satellite propagation and processing delays and the delay introduced by the terrestrial B-ISDN.

The MAC packet slot period has been chosen to support a 32 kbit/s CBR stream and corresponds to one frame unit of 384 un-coded information bits every uplink frame. This results in a frame period of 11.9 ms to transmit 84 ATM cells per second using AAL5. There are 64 MAC packet slots to support 2.048 Mbit/s of traffic per spot-beam on the uplink. The actual uplink transmission rate is higher due to ATM and MAC layer overheads.

3.3. Cell level. The ATM network services are precisely defined in 20 and they are: CBR, VBR, ABR and UBR. WATM network should offer these services with the required QoS parameter values as shown in Table

1 11. These parameters correspond to the ATM cell level QoS parameters and they were defined from the point of view of the user application.

Table 1 QoS parameters for specific ATM services

	CBR, rt-VBR	nrt-VBR	ABR
CLR	$10^{-7} - 10^{-4}$	$10^{-9} - 10^{-6}$	$10^{-9} - 10^{-4}$
CDV	< 1 ms	not specified	not specified
max CTD	2s – 10ms	10s – 500ms	not specified
PCR	10Mbps – 32kbps		10Mbps – 9.6kbps
SCR	6Mbps – 32kbps	6Mbps – 32kbps	
BT	5Mbps – 2kbps	1Mbps – 1kbps	

3.3.1 Evaluation of CBR, VBR and UBR services. In this section we summarise the results presented in 25, 26 concerning the performances of the ATM network services in the fixed wireless environment (reference configuration #1). The assumed configuration is a bottleneck topology. The tested connections of different type were established between two terminals, each of them connected to the network by a RAP (Radio Access Point) unit. Furthermore is assumed that the radio access to the network is governed by the MEDIAN protocol 27 and transmission errors occur independently.

Transmission errors. The radio channels are usually characterised by relatively high BER value, even up to 10^{-2} 28. The situation is going worst when the terminal is on the move and then the transmission conditions can be temporary worst due to fading and shadowing 28, 29, 30. As a consequence, even that a powerful protection mechanism is used the cells could be lost even in bursts.

MAC protocol. A number of new MAC protocols for the purpose of the wireless ATM network was recently submitted, among them the most recognised are MASCARA 28, SAMBA 31 and MEDIAN 27. These protocols were designed to operate under TDMA (*Time Division Multiple Access*) scheme, where the time slot allocation for a given connection is made dynamically. In the MEDIAN, the time slots are assigned to the active connections on the basis of the polling information and the connection priority. The priority strictly corresponds to the maximum permitted cell waiting time (Δ_{max}). A cell is discarded when it waits longer than Δ_{max}. Notice that a cell from higher priority connections could be served after serving the cell of lower priority. Therefore, this protocol does not support the possibility to reserve a fixed number of slots in consecutive transmission frames, what is extremely required e.g. for CBR connection.

Exemplary numerical results (partially verified by measurements) were obtained by using OPNET package environment. Based on them one can conclude as follows:

CBR service. Additional mechanisms should be implemented in order to support CBR service. One solution is to adjust the Δ_{max} parameter in the MEDIAN protocol. By setting larger value of the Δ_{max} one can expect larger CDV, but the cell loses can be avoided. On the contrary, low value of the Δ_{max} leads to low CDV, but cell loss ratio increases. Another possible solution is to keep large Δ_{max} value and to provide a buffer at the connection end point. This buffer introduces fixed delay but the cells are delivered to the destination with constant rate and very low CDV. In 32, 33, 34 some propositions for a playback buffer management scheme for the CBR service in the fixed ATM network were submitted. As it was expected these solutions fail in the case of WATM network due to high bit error rate and large CDV. In 26 we propose appropriate modifications for the playback buffer mechanism. The proposed enhancements were related to the strategy of inserting dummy cells instead of lost ones and extension K times cell numbering scheme by assigning the same number to K consecutive cells.

VBR service. Satisfying the requirements from Table 1 it seems to be difficult to reach. The features of the MEDIAN protocol, which introduces too large delay, mainly cause this. However, from the application viewpoint, these relatively large CTD values (about 23 ms) are still acceptable (see Table 2 35). Anyway, a similar mechanism, like it was proposed for the CBR service, to reduce CDV, is extremely desirable.

Table 2 Allowed values of max CTD for various types of applications

Application	Bandwidth	Max CTD
Voice	8 - 16 kbps	150 ms
Videotelephony	64 kbps - 2 Mbps	350 ms
Videoconference	128 kbps - 14 Mbps	350 ms
Transaction processing	64 kbps - 5 Mbps	3 s
Telescript session	4 - 5.6 Mbps	250 ms

TCP over UBR Service. The TCP greedy source was used to evaluate the effectiveness of UBR service. The gathered cell transfer characteristics correspond to throughput, window size and RTT. In the running experiments the bit error rate was $7.5 * 10^{-3}$ (for higher bit error rate the TCP segment lost probability was too high). The cell loss ratio (CLR) for this BER was

10^{-2}. The obtained results say than even assuming so heavy conditions, the effectiveness of cell transfer is still acceptable.

3.3.2 ABR flow-control. Connections of the ABR service class are designed to use the remaining capacity of an ATM network after the CBR and VBR service-class connections have received their current cell rate. To avoid congestion in an ATM network and to use the whole capacity efficiently, the source of every ABR connection has to be informed regularly about this remaining capacity. For this purpose, the ATM Forum has standardized rate-based ABR flow-control algorithms.

Considering the high-speed data transfer by the ABR service class in a wireless ATM (WATM) network, the impact of forward and backward hard handover protocols on the performance of the two rate-based ABR flow-control schemes Explicit Rate Indication for Congestion Avoidance improvement (ERICA+) and Fuzzy Explicit Rate Marking Adaptation (FERMA) has been investigated (cf. 36, 37, 38. For this purpose, the ABR service class that is specified in the ATM forum document Traffic Management Specification 4.0 is used (cf. 20). The underlying WATM architecture and the used protocol stack of this study are derived from NEC's prototype implementation WATMnet and the ATM forum documents (cf. 39, 40, 41). It includes a simplified version of the TDMA/TDD structure and the associated wireless MAC protocol of this system. Using a model of a basic client-server scenario in a wireless ATM network, the impact of the error-prone wireless communication channel and of mobility-management techniques determined by backward and forward hard handover protocols has been studied (cf. 39, 42, 43, 44, 45).

3.4. Handover protocols. Regarding data communication in a WATM network the connections must be permanently maintained during the communication phase since a mobile terminal (MT) moves within a certain coverage area of a base station (BS) and may cross its boundary (cf. 39, 42, 43, 44, 45). In this case, the connections must be handed over to a new transmission cell whereby a new radio channel is seized and the QoS requirements of the corresponding virtual channels (VCs) must be satisfied in addition to the existing ones within the new cell. The corresponding hard handover protocol has to guarantee the sequence integrity and loss-free delivery of the ATM cells during this transition process. Moreover, interworking with the rate-based ABR flow-control algorithms is required to guarantee the effectiveness of the approach (see 36, 38 and references therein, also section 3.3.2).

To achieve these goals, an improved backward hard handover protocol has been developed for handovers between the transmission cells of the current BS (CBS) and a new BS (NBS) of the MT, called better hard handover (see Fig.

19 for a simplified error- and loss-free message transfer without acknowledgements - cf. 36).

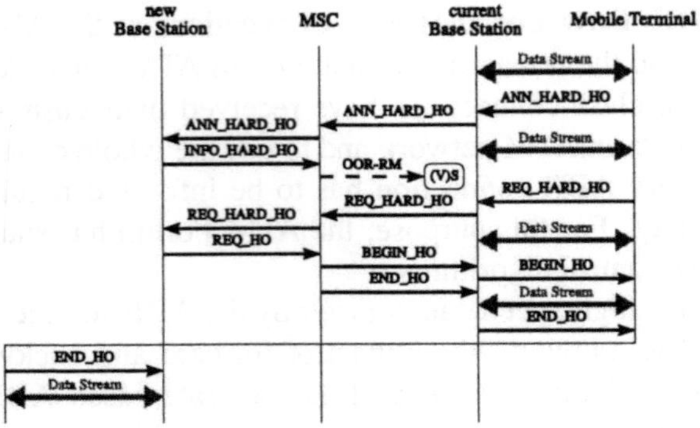

Figure 19 Better Hard Handover

Derived from a congestion-awareness concept the new handover protocol is a minimal enhancement of a standard hard handover protocol. It includes only a few new signaling messages using special RM cells. Independent of handover events they are exchanged between the base stations and the mobile switching center (MSC) which is an ATM switch with mobility support enhancements for the wireless part of the network. Whenever there is a substantial change of the ABR bandwidth or an alteration in the number of active ABR connections in the transmission area of a BS the BS informs the MSC immediately by sending an MSC information message with the current ABR target cell rate TCR, the number of ABR connections and the sum of their minimum cell rates. This very low additional bandwidth requirement in the path between each BS and the MSC gives the MSC the knowledge about the current possible explicit rate (ER) in each BS transmission area.

The improved backward hard handover is combined with forward handover as a fallback procedure in the case of a too short announcement period before a handover. This means that normally every mobile terminal announces its expected new handover t_a milliseconds before the real handover event occurs. If there is not enough time for such an announcement, the mobile terminal has to perform a forward handover (see Fig. 20, cf. 36).

4. WIRELESS LANS

In ad-hoc WLANs, where every station has a similar functionality, Carrier Sense Multiple Access (CSMA) based protocols may be advantageous because of their simplicity and because no station is compelled to assume special functions. Recently proposed standards such as the IEEE 802.11 or HIPERLAN I use variations of the basic CSMA scheme.

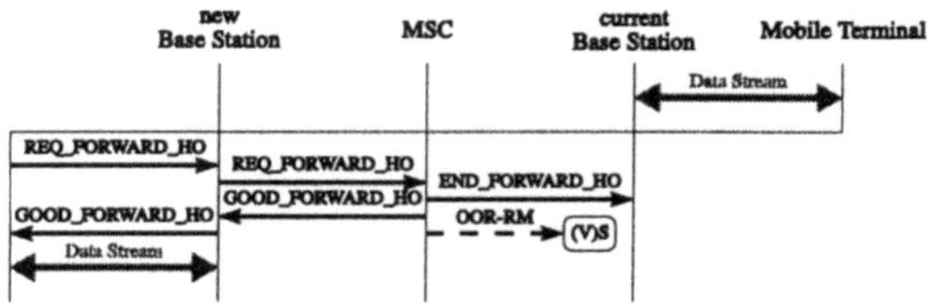

Figure 20 Forward Handover

However, if we are to integrate delay-sensitive and delay-insensitive traffic in the same WLAN, we must keep end-to-end delay and delay variations below certain bounds for the delay-sensitive traffic. Unfortunately, it is well known that in general CSMA protocols suffer from large access delay quantiles (the basic CSMA algorithm is actually unstable) due to the existence of packet collisions.

This section describes FIFO by Sets CSMA (FS-CSMA) 46, a MAC protocol for high-speed ad-hoc WLANs. FS-CSMA is a Collision Resolution Algorithm that builds groups (Transmission Sets) with the packets arrived during fixed-length time intervals. Packets belonging to the same Transmission Set are transmitted using CSMA with random backoff whereas the different Transmission Sets are served following a FIFO discipline.

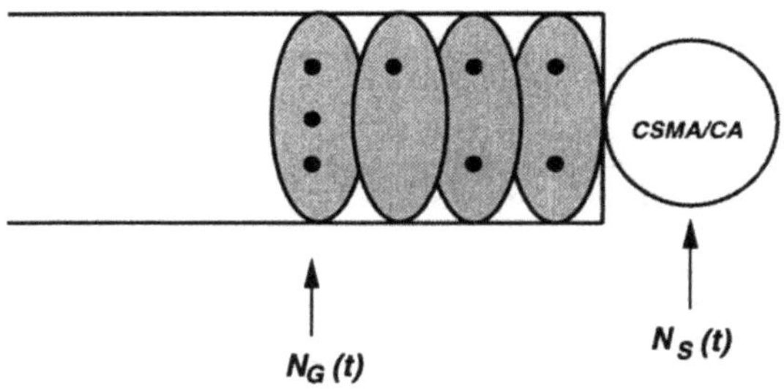

Figure 21 FS-CSMA as a distributed FIFO queueing system

FS-CSMA generates a collection of (possibly empty) sets called Transmission Sets (TSs) $\{S_k\}$, each of them containing the packets arrived during successive disjoint grouping intervals of constant length W slots. These Transmission Sets are served (i.e. their included packets are transmitted using CSMA) with a FIFO discipline, so that S_k is served before S_j if $k < j$. If every S_k had no more than one packet to be transmitted we would be facing a FIFO-MAC algorithm, with packets being sent in order of arrival.

The number of packets per TS is closely related to the integer W, which determines the length of the grouping interval, and to the network load ρ. The smaller W or ρ are the more the system behaves like a FIFO system. On the contrary, the protocol will perform as pure CSMA if *W* is made large enough. The value of *W* must be chosen as a trade-off between the desired properties of a FIFO scheduling and the overhead introduced by the algorithm.

An FS-CSMA ad-hoc network may be seen as a distributed FIFO queueing system (see Fig. 21). The elements of this distributed queue are the Transmission Sets and their service consists of the successful transmission via CSMA with random backoff of all the packets included in the Transmission Set.

REFERENCES

[1] H. Nakamura. H. Tsuboya and A. Nakajima. Applying ATM to Mobile Infrastructure Networks. *IEEE Communication Magazine*. 36(1): 67-73.1998

[2] N. Gerlich. The Impact of Wireless Communication on Broadband Network Architecture and Performance. *Proceedings of the COST 257 Mid-Term Seminar*. Jan.1999.

[3] T. Ojanpera and R. Prasad. Wideband CDMA for third generation mobile communications.
Artech House Publishers. 1998.

[4] P. Tran-Gia and N. Gerlich. Impact of Customer Clustering on Mobile Network Performance. COST 257 document. 1997.[257TD(97)07]

[5] T. Leskien K. Tutschku. and P. Tran-Gia. Traffic estimation and characterization for the design of mobile communication networks. COST 257 document. 1997. [257TD(97)22].

[6] M.-A. Remiche. Efficiency of an iphp^3 illustrated through a model in cellular networks. COST 257 document. 1998. [257TD(98)13].

[7] M.-A. Remiche. Impact of non-stationary users distribution on the outage probability measure in cdma wireless system. [257TD(99)01].

[8] Tran-Gia and M. Mandjes. Modeling of customer retrial phenomenon in cellular mobile networks.*IEEE Journal on Selected Areas of Communications*, 15(08):1406-1414, Oct. 1997. [257TD(97)06].

[9] K. Elsayed N. Gerlich, P. Tran-Gia. and N. Jain. Performance analysis of a link carrying capacity in cdma systems. COST 257 document. 1997. [257TD(97)22].

[10] N. Gerlich and M. Menth. The performance of aal-2 carrying cdma voice traffic. COST 257 document. 1998. [257TD(98)32].

[11] L. Dellaverson. Wireless atm requirements specification. ATM Forum. Draft Specification. 01.03. Feb.1999, ATM Forum, Feb. 1999.

[12] B. Rajagopalan. Draft wireless atm capability set 1 specification. Draft Specification. 01.12, ATM Forum, Sep. 1999.

[13] G. Carle and E. Dorner. Assessment of edge devices with error control mechanisms for ATM networks with wireless local loop. Personal Wireless Communications. Aachener Beitrage zur Informatik. Verlag Augustinus Buchhandlung. 1996.

[14] D. Ginsberg. ATM solutions for enterprise interworking. Addison-Wesley, New York, 1996.

[15] C. H. C. Leung and et al. The throughput efficiency of a go-back-n arq scheme under Markov and related error structures. IEEE Trans. on Comm., 36(2):231-234, 1988.

[16] D. L. Lu and J. F. Chang. Performance of arq protocols in nonindependent channel errors. IEEE Trans. on Comm., 41(4):721–730, 1993.

[17] T. N. Saadawi and et al. Fundamentals of Telecommunication Networks. John Wiley & Sons. 1994. New York.

[18] H. Ohta and T. Kitami. A cell recovery method using fec in atm networks. IEEE Journal on Selected Areas in Communications. 9(9):1471–1482, 1991.

[19] G.R. Pieris and G.H. Sasaki. Performance of the go-back-∞ protocol under correlated packet losses. IEEE Trans. on Comm., 41(5):660–663, 1993.

[20] N. Giroux. Traffic management specification version 4.0. Specification. 4.0, ATM-Forum April. 1996.

[21] A.K. Elhakeem S. Bohm. and V.K.M. Murthy. Analysis of a movable boundary random/dama accessing technique for future integrated services satellites. IEEE Globecom, pages 1283-1289, 1993.

[22] N. Celandroni and E. Ferro. The foda-tdma satellite access scheme: Presentation, study of the system and results. IEEE Trans. Comms. 39(12): 1823-1831, 1991.

[23] M. Tondriaux T. Zein, G. Maral. and D. Seret. A dynamic allocation protocol for a satellite network integrated with b-isdn. Proc. of 2nd ECSC, pages 15-20, 1991.

[24] ITU-T. B-ISDN ATM layer cell transfer performance. Recommendation, I.356 ITU-T. 1996

[25] W. Burakowski A. Bak, A. Beben and Z. Kopertowski. On quality of watm network services. COST 257 document, 1999. [257TD(99)11].

[26] Beben and W. Burakowski. On improving cbr service playback buffer mechanisms in watm network. COST 257 document. 1999. [257TD(99)12].

[27] MEDIAN Partners. Median final system design. Interim project deliverable, 1996.

[28] G. Marmigcre L. Merakos F. Bauchot, S. Decrauzat. and N. Passas. Mascara, a mac protocol for wireless. Interim project deliverable.

[29] B.Sklar. Rayleigh fading channels in mobile digital communication systems. part I: Characterization, IEEE Communication Magazine, September1997.

[30] B.Sklar. Rayleigh fading channels in mobile digital communication systems. part II: Mitigation. IEEE Communication Magazine, September 1997.

[31] ACTS 2004 Program SAMBA. Samba - System for advanced mobile broadband application. Interim project deliverable.

[32] B. Bharucha M. Noorchashm. and G. Wetzel. Buffer design for constant bit rate services in presence of cell delay variation. ATM Forum document. af95-1454, ATM Forum, 1995.

[33] J. Virtamo (Eds.) J. Roberts. U. Mocci. Broadband network teletraffic. Performance evaluation and design of broadband multiservice networks, Final report of action COST 242, 1996.

[34] ITU-T. B-Isdn atm adaptation layer (aal) specification. Recommendation. I.363.

[35] M. Niemi. Application requirements for watm. ATM Forum document. af96-1058, ATM Forum, 1996.

[36] U.R. Krieger and M. Savoric. Performance evaluation of handover protocols for data communication in a wireless atm network. Proc of ITC, 1999.

[37] U. R. Krieger and M. Savoric. The Performance of abr flow-control algorithms and flow control protocols in a wireless atm network. COST 257 document. 1998.[257TD(98)31].

[38] Udo R. Krieger and M. Savoric. Performance evaluation of handover protocols for data communication in a wireless atm network. COST 257 document, 1998. [257TD54].

[39] K.Rauhala. Baseline text for wireless atm specifications. Draft Specification. btd-watm-01.07, ATM Forum, April 1998.

[40] J. Porter and et al. The orl radio atm system, architecture and implementation. Intermin report, Olivetti research Ltd, January 1996.

[41] D.C. Cox. Wireless personal communications: What is it?. IEEE Personal Communications, pages 20–35, April. 1995.

[42] K. Y. Eng E. Ayanoglu. and M. J. Karol. Wireless atm: Limits, challenges and proposals. IEEE Personal Communications, pages 18–34. August1996..

[43] G.P. Pollini. Trends in handover design. IEEE Communications Magazine, pages 82–89, March. 1996.

[44] R. Ramjee and et al. Performance evaluation of connection rerouting schemes for atm-based wireless networks. IEEE/ACM Transactions on Networking, 6(3), 1998.

[45] C.K. Toh. Wireless ATM and Ad-Hoc Networks. Kluwer Academic Publishers, Boston 1997.

[46] Vázquez-Cortizo. D. and Garcia. A collision resolution algorithm for ad-hoc wireless lan, Broadband Communications, pages 119–130, 1998.

[47] Ch. Bulloch. Ka-band Abroad: The world rises to the challenge, Via Satellite, March 1997.

[48] R. Fernandez. The Ka-band quest continues, Via Satellite, March 1997.

[49] E.F. Fitzpatrick. Hughes Spaceway: Wireless interactive broadband service, Proc. of Second Ka-Band Utilization Conference, September 24–26, 1996.

[50] T. Ors. S. Sun and B.G Evans. A meshed vsat satellite network architecture using an on-board atm switch, IEEE IPCCC, 208–215, February 1997.

[51] W.R. Stevens. TCP/IP illustrated, Addison-Wesley. 1995.

[52] B. Jabbari. G. Colombo. A. Nakajima and J. Kulkarni. Network issues for wireless communications, IEEE Communications Magazine, pages 88–98, January 1995.

[53] D. Raychaudhuri and et al. WATMnet: A prototype wireless atm system for multimedia personal communication, IEEE Journal on Selected Areas in Communications, 15(1): 83–95, 1997

[54] R. Jain and et al. ERICA Switch Algorithm: A complete description, ATM Forum document, af96-1172, ATM Forum August 1996.

[55] U. R. Krieger and M. Savoric. Adaptation of the fuzzy explicit rate marking to abr flow-control in a wireless atm network, Proc. of MMB, September 1999.

End-to-End and Redirection Delays in IP Based Mobility

Jon-Olov Vatn
Department of Teleinformatics Royal Institute of Technology
Electrum 204 S-164 40 Kista, Sweden, Email: vatn@it.kth.se

Key words: IP, mobility, end-to-end delay, handover, routing

Abstract: Using the Internet as an infrastructure for mobile, real-time communication is an attractive goal as well as a challenging task. The non-optimal routing inherent in several proposed IP mobility schemes makes it harder to meet the requirement of low end-to-end delay. Mobile users may also perceive decreased performance when a handover between to access points is performed. In this study, common IP based mobility support schemes are evaluated according to their impact on the end-to-end delay and their IP level handover performance. Of the schemes considered, Mobile IPv6 [10] shows the best characteristics. Mobile IPv4 with route optimization [11] is also promising, however, some enhancements are suggested.

1. INTRODUCTION

Real-time interactive services like telephony have high requirements on low end-to-end delay (T_{ee}) and low delay variation. Providing this kind of service over the Internet leads to several challenging problems, due to the delay variations inherent in datagram networks. The delay variations can be addressed by buffering at the receiver, but the requirement on low maximum end-to-end delay ($T_{ee,max}$) is still an issue ($T_{ee,max}$ is often considered to be around 150 ms, but may be stretched up to 400 ms for trained users[5]).

When extending these real-time services to include mobile users, it will be even harder to meet the requirement of low T_{ee}, since packets may not take the shortest path between the communicating entities due to non-optimal routing, and it is difficult to guarantee an acceptable quality of service in all the locations that the user may move to. Also, mobile users may perceive reduced performance when moving between different access points (APs), i.e., performing a handover. Packets *in-flight* to the mobile user's previous point of attachment, may be lost or unacceptably delayed. The time interval when users may perceive reduced performance due to a handover will depend on factors such as movement detection delay, time to acquire a new care-of IP address [13, 14], and the time it takes to redirect (T_{rd}) data to this new location.

The objective of this paper is to compare a set of proposed mobility support schemes and evaluate their performance when the mobile user is visiting a foreign IP subnet (by looking at T_{ee}), and when it performs a handover between subnets (by looking at T_{rd}). In section 2 we cover the basic concepts of IP mobility. In section 3, the different mobility support schemes are presented and analyzed using symbolic expressions for T_{ee} and T_{rd}. These results are summarized and discussed in section 4. Finally, section 5 concerns future work.

2. IP BASED MOBILITY

Each node node in the Internet is identified by one (or more) IP address(es), and the IP addresses of nodes on the same IP subnet have a common prefix. This makes routing scalable, since routers only have to keep track of networks, not individual hosts. However, this implies that a node that leaves one IP subnet to attach to another will have to acquire a new address corresponding to the new subnet. Furthermore, the node needs to be identified by same address as before to be able to keep its ongoing higher level connections.

Therefore, Mobile IPv4 (MIPv4)[8] lets the mobile nodes (MNs) be associated with two IP addresses: one *home address* belonging to a home network, and a *care-of address* (COA) associated with the visited subnet. When the MN is away from home, a host on the home network, a home agent (HA), intercepts packets destined for the MN and *tunnels*[9] them to the MN's current COA. This tunnel can end at the MN itself or at a dedicated host, a foreign agent (FA), on the visited subnet. In the former case we call the COA a *co-located* COA and in the latter case we have a *foreign agent* COA. When a MN moves to a new subnet, it needs to inform its HA about its new COA, so that the HA can redirect packets accordingly.

For traffic going in the other direction (upstream), MIPv4 offers two possibilities: the MN can either send data directly to the correspondent node (CN), or tunnel the packets back to its HA, which in turns routes them towards the CN. These two routing schemes are called *triangular routing* and *reverse tunneling*, see Figure 1.

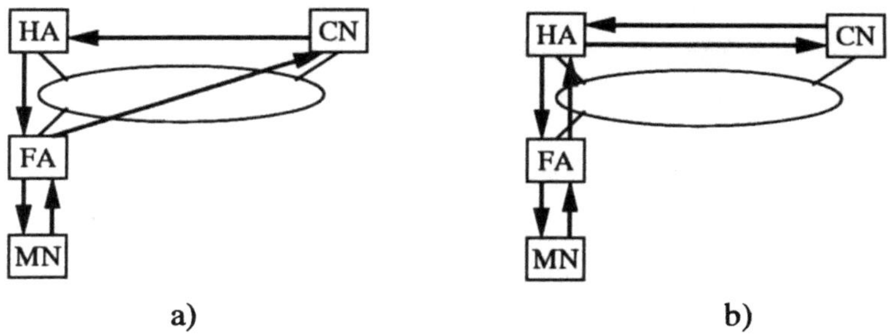

Figure 1 Mobile IPv4 with (a) triangular routing and (b) reverse tunneling.

In the next section, we will analyze MIPv4 and a set of other proposed IP mobility support schemes and evaluate them with respect to their effect on the end-to-end delay (T_{ee}) and the redirection delay (T_{rd}). Since these metrics may differ for the downstream (towards the MN) and the upstream (towards the CN) flow, we define these delays separately.

Definition 1 Downstream end-to-end delay ($T_{ee,do}$): The time elapsed from when a CN sends a packet until it arrives at the MN. □

Definition 2 Upstream end-to-end delay ($T_{ee,up}$): The time elapsed from when a MN sends a packet until it arrives at the CN. □

Definition 3 Downstream redirection delay ($T_{rd,do}$): The amount of time when packets are lost (or delayed so that $T_{ee,do} > T_{ee,max}$) after a MN has acquired a new COA, because the packets were still directed to the old COA. □

Definition 4 Upstream redirection delay ($T_{rd,up}$): The amount of time a MN is hindered from sending data to its CN(s) via the new subnet after the MN has acquired a new COA. □

The definitions of $T_{ee,do}$ and $T_{ee,up}$ are quite straightforward, however, one should note that we do not consider the delay associated with packetization at the sender. If we compare these two metrics for the setup in Figure 1, one can expect that $T_{ee,do}$ will be equal for triangular routing and reverse tunneling, while $T_{ee,up}$ will be lower for triangular routing.

Regarding T_{rd}, the definitions need some further explanation. We assume that the MN is roaming in area with overlapping cells, but that the MN can only attach to one AP at a time. Then $T_{rd,do}$ represents the period of time when packets are lost, if the time to shift between the APs and the time to acquire a new COA were both zero. For the upstream traffic the situation is different, since the MN itself could be seen as the point of redirection. Nevertheless, for some of the mobility support schemes the MN may be hindered from sending data to the CN for a certain amount of time $T_{rd,up}$.

3. ANALYSIS OF IP MOBILITY SCHEMES

There exist several proposals for alternatives or extensions to the basic MIPv4 scheme introduced in section 2. The additional schemes we have considered are referred to as Mobile IPv4 with route optimization [11], Mobile IPv4 with Hierarchical Foreign Agents[7] and Mobile IPv6 (MIPv6)[10]. The first two contain extensions to MIPv4 that reduce T_{ee} and T_{rd} respectively. MIPv6 is covered since it will become important when IPv6 deployment takes off.

We will present symbolic expressions for the performance metrics defined in section 2, for each of these schemes. We restrict our analysis to the case were *both the MN and the CN are mobile* and we let them use *the same mobility support scheme.* The only difference between the MN and the CN is that the MN is performing a handover, while the CN is staying at a certain location during this period of time.

Sections 3.1-3.4 concern different IPv4 mobility support schemes based on the use of foreign agents, while section 3.5 treats aspects of using co-located COAs in MIPv4. Section 3.6 concerns mobility support in IPv6.

3.1. MIPV4 WITH TRIANGULAR ROUTING

As described in section 2, MIPv4 offers two possible routing schemes, triangular routing and reverse tunneling. Here we will present the delays in triangular routing, while reverse tunneling is covered in the next section.

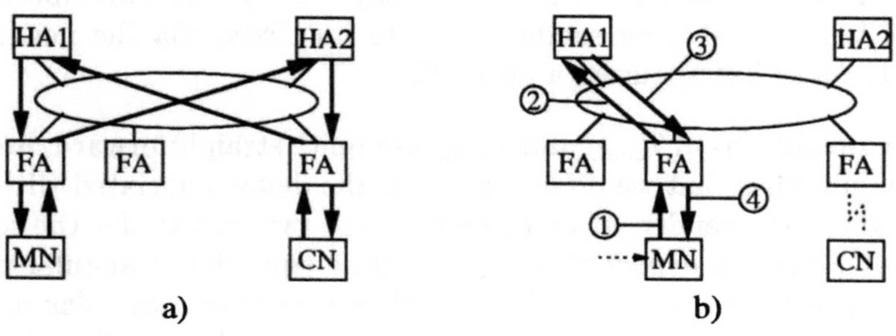

Figure 2 MIPv4 with triangular routing: (a) data paths, and (b) signaling paths.

The data and signaling paths for MIPv4 with triangular routing are shown in Figure 2. Since the CN is also mobile (with home agent HA2), the data path takes the form of a bow-tie(⋈) and not a triangle as in Figure 1. Symbolic expressions for $T_{ee,do}$ and $T_{ee,up}$ are shown in equation 1 and 2.

$$T_{ee,do} = T_{cn->ha1} + T_{ha1->mn} \quad (1)$$
$$T_{ee,up} = T_{mn->ha2} + T_{ha2->cn} \quad (2)$$

$T_{cn->ha1}$ is the delay for packets to travel from the CN to the HA of MN, $T_{ha1->mn}$ is the delay from the HA to MN and so on. Processing delays at intermediate nodes, e.g., HAs and FAs, are not stated explicitly. Instead we emphasize the delay related to the path the packets will traverse. We do not claim that processing delay is irrelevant, however, its impact decreases with improved hardware performance.

To redirect the downstream traffic, the MN sends a *Binding Update* (*BinUp*) to its HA via the new FA. The HA responds with a *Binding Acknowledgment* (*BinAck*) to acknowledge a successful (or unsuccessful) binding update and redirects the packets to the new COA. Packets lost after the MN has acquired the new COA are the ones ***in-flight*** between the HA and the previous COA plus the ones arriving at HA while the ***BinUp*** travels from the MN to the HA. Thus, $T_{rd,do}$ will be the sum of the network delay between the HA and the MN's old and new point of attachment (denoted $T_{ha1->mn,old}$ and $T_{mn->ha1}$ respectively).

The time to redirect the upstream data flow depends on the round-trip time from the MN to its HA, i.e, the MN should wait for a (positive) *BinAck* in response to the *BinUp*[8] before it updates its routing table in accordance with the new subnet. Symbolic expressions for $T_{rd,do}$ and $T_{rd,up}$ are given in equation 3 and 4.

$$T_{rd,do} = T_{ha1->mn,old} + T_{mn->ha1} \quad (3)$$
$$T_{rd,up} = T_{mn->ha1} + T_{ha1->mn} \quad (4)$$

3.2. MIPV4 WITH REVERSE TUNNELING

In Mobile IPv4 with triangular routing, the MN uses its home IP address as source address, even if it is currently attached to another network. This may lead to problems, since security conscious routers may drop packets, which do not have a topologically correct source address [3]. One way to address this problem is to let the FA tunnel the packets from the MN to the CN via the HA, as shown in Figure 3. These packets will be able to pass ingress filtering routers, since the source address of the "outer" IP header will be the IP address of the FA. This method is called reverse tunneling[6], since both downstream and upstream traffic are tunneled between the HA and the FA. (Reverse tunneling can also be useful if the MN sends multicast packets, since multicast routing

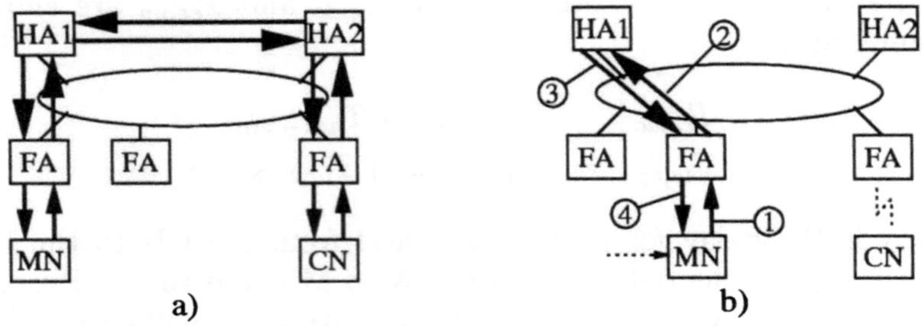

Figure 3 **MIPv4 with reverse tunnels: (a) data paths, and (b) signaling paths.**

protocols based on *reverse path forwarding* also depend on topologically correct source addresses.) Symbolic expressions for $T_{ee,do}$ and $T_{ee,up}$ are given in equations 5 and 6.

$$T_{ee,do} = T_{cn->ha2} + T_{ha2->ha1} + T_{ha1->mn} \quad (5)$$
$$T_{ee,up} = T_{mn->ha1} + T_{ha1->ha2} + T_{ha2->cn} \quad (6)$$

The tunnel from the FA to the HA will not be available before the FA receives a positive *BinAck* from the HA, thus sending data from the MN before receiving the *BinAck* makes little sense, since the FA is likely to drop them. Hence, if reverse tunneling is used, the MN will not be able to send data packets via the new FA before the *BinAck* has been received. $T_{rd,do}$ and $T_{rd,up}$ will be the same as for MIPv4 with triangular routing, see equation 3 and 4.

3.3. MIPV4 WITH ROUTE OPTIMIZATION

To reduce the end-to-end delays present in MIPv4, there is an IETF draft[11] proposing a scheme for route optimization. This could be seen as an extension to the MIPv4 protocol, where CNs can tunnel packets directly to the MN's current COA, instead of routing them via the HA. When a HA, that supports route optimization, intercepts data destined for one of its MNs, it will tunnel the packets from the CN to the MN (as before), but also send a *BinUp* to the CN to inform it about the MN's current COA. Thus, a CN will send the first (few) packet(s) via the HA, but is then able tunnel the packets directly to the MN's current location, see Figure 4a). The MN will also tunnel packets (upstream) to the CN, using its home address source address in the "outer" and the "inner" header. $T_{ee,do}$ and $T_{ee,up}$ will be low, see equations 7 and 8.

$$T_{ee,do} = T_{cn->mn} \quad (7)$$
$$T_{ee,up} = T_{mn->cn} \quad (8)$$

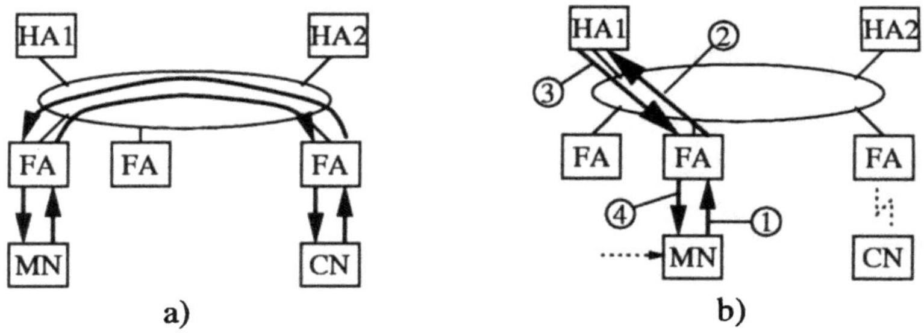

Figure 4 MIPv4 with route optimization: (a) data paths, and (b) signaling paths.

To avoid losing all packets ***in-flight*** between the HA and the old FA, the route optimization draft also includes support for *smooth handover*. The MN can inform its old FA where it should forward packets it receives after the MN has moved. $T_{ee,do}$ for these packets will be

$$T_{ee,do} = T_{cn->fa,old} + T_{fa,old->mn} \quad (smooth) \tag{9}$$

The handover procedure differs somewhat from the schemes presented earlier. The MN will inform its HA (and possibly its old FA) about its new COA. The HA will then send a ***BinUp*** to inform the CNs, so that they can redirect the downstream packets, see Figure 4b). (***BinUps*** can also be triggered if the HA is notified about CNs with stale mobility bindings by the old FA.) Due to the smooth handover feature, data ***in-flight*** can be redirected by the old FA, thus decreasing $T_{rd,do}$ if the new and old FA are relatively close and if $T_{ee,do}$ in equation 9 is less than $T_{ee,max}$. Smooth handover is likely to have most impact when the handover is performed between IP subnets within the same administrative domain (AD), i.e., an ***intra-AD*** handover. $T_{ee,up}$ ought to be the same as for the base MIPv4 scheme, see section 3.1. Hence, the redirection delays will be as follows:

$$T_{rd,do} = \begin{cases} T_{mn->fa,old} & \text{intra-AD} \\ T_{cn->mn,old} + T_{mn->ha1} + & \\ T_{ha1->cn} & \text{cross-AD} \end{cases} \tag{10}$$

$$T_{rd,up} = T_{mn->ha1} + T_{ha1->mn} \tag{11}$$

Equation 11 assumes that HA1 has cached a binding to the CN's current location. This assumption is reasonable since the CN will regularly contact HA1 to check the validity of the binding it has cached for MN.

3.4. MIPV4 WITH HIERARCHICAL FOREIGN AGENTS

Perkins introduces a mobility support architecture with multiple levels of redirection[7]. The idea is that handovers performed within an AD can be handled locally. This will both reduce the amount of signaling sent over the backbone and lead to faster handovers. FAs are arranged in a hierarchy, where packets destined for the MN arrive at some top level FA. Packets will then be redirected through multiple tunnels until they reach the FA closest to the MN, see Figure 5a). When a MN performs a handover, the *BinUp* only needs to travel to the FA that constitutes the lowest common node (the branching point) in the FA hierarchy. If there is no common FA, the *BinUp* will have to go all the way to the HA. As a hierarchy of FAs are not likely to span multiple ADs, only *intra-AD* handovers are likely to benefit from this architecture.

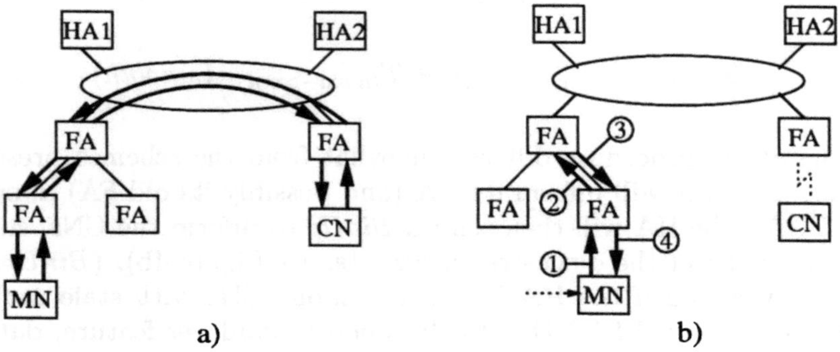

Figure 5 MIPv4 with hierarchical foreign agents: (a) data paths, and (b) signaling paths.

This scheme was designed as an extension of MIPv4; however, schemes based on hierarchical FAs could be combined with several other schemes for *cross-AD* handover. If we consider the case where hierarchical FAs are combined with the route optimization scheme of section 3.3 (skipping the smooth handover feature for simplicity) T_{ee} and T_{rd} will be as follows:

$$T_{ee,do} = T_{cn->mn} \tag{12}$$

$$T_{ee,up} = T_{mn->cn} \tag{13}$$

$$T_{rd,do} = \begin{cases} T_{fa,bra->mn,old} + T_{mn->fa,bra} & \text{intra-AD} \\ T_{cn->mn,old} + T_{mn->ha1} + T_{ha1->cn} & \text{cross-AD} \end{cases} \tag{14}$$

$$T_{rd,up} = \begin{cases} T_{mn->fa,bra} + T_{fa,bra->mn} & \text{intra-AD} \\ T_{mn->ha1} + T_{ha1->mn} & \text{cross-AD} \end{cases} \tag{15}$$

3.5. MIPV4 WITH CO-LOCATED CARE-OF ADDRESSES

Instead of relying on FAs, the MN could acquire a COA of its own, e.g., by use of DHCP[2]. This kind of COA, called a *co-located* COA, could be used instead of *foreign agent* COAs in the schemes already presented[1].

The only significant difference in the metrics we are studying concerns $T_{rd,up}$. Once the MN has acquired a COA on a new subnet, it should be able to send data to its CNs immediately, since there is no need to wait for a positive *BinAck*. Thus, $T_{rd,up}$ is equal to zero for the triangular routing, reverse tunneling and route optimization schemes if co-located COAs are used.

$$T_{rd,up} \quad = \quad 0 \quad \text{(all MIPv4 schemes)} \tag{16}$$

The route optimization draft[11] does not consider co-located COAs, but it should work well except for two features that rely on the existence of FAs; smooth handover and stale mobility binding detection, see section 3.3. Since there is no smooth handover support, $T_{rd,do}$ will be as long for intra-AD as for cross-AD handovers when using MIPv4 with route optimization, see equation 17 (compare with equation 10).

$$T_{rd,do} \quad = \quad T_{mn->ha1} + T_{ha1->cn} + T_{cn->mn,old} \tag{17}$$

3.6. MOBILITY SUPPORT IN IPV6

Work is in progress to design mobility support in IPv6 (MIPv6)[10]. The major difference with respect to MIPv4 is that there are no FAs in MIPv6, i.e., only co-located COAs are used. To get a COA on a new IPv6 subnet should be relatively easy, as the MN has the possibility to use stateless[12] and stateful[1] IPv6 address autoconfiguration.

In MIPv6 a CN is able to send data directly to the MN so $T_{ee,do}$ and $T_{ee,up}$ will be the same as for route optimization in MIPv4, see equations 7 and 8.

When the MN has moved to a new IPv6 subnet and acquired a COA on this subnet, it will send a *BinUp* message[2] to its HA to register this COA. The MN can also send a *BinUp* to its CN(s), and also to a router[3]

[1]To use co-located COAs in a hierarchical FA setup would be possible if the MN's are informed about the IP address of the FA above it in the hierarchy. This could be accomplished, e.g., by adding a hierarchical FA option in DHCP.

[2]It does not have to be a separate IPv6 packet, as a binding update destination option can be included in any existing packet being sent to the same destination.

[3]This kind of router needs special support for smooth handover.

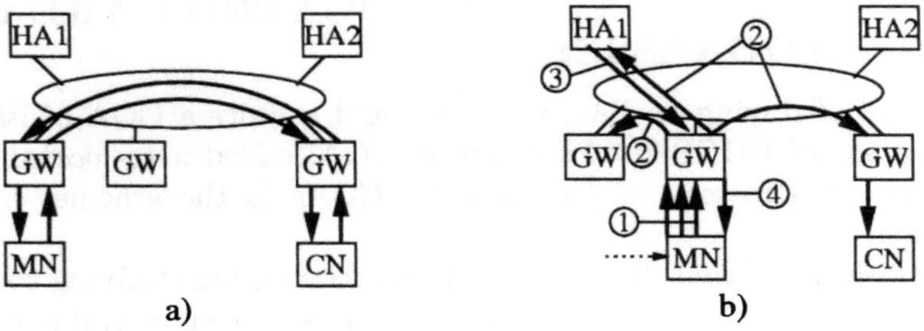

Figure 6 MIPv6 with route optimization: (a) data paths, and (b) signaling paths.

on its previous subnet (for smooth handover). Since the MN can send a *BinUp* directly to its CN, $T_{rd,do}$ will be shorter, as compared to route optimization in MIPv4 (see Equation 10). $T_{rd,up}$ will be approximately zero, since the MN can redirect its outgoing flow immediately.

$$T_{rd,do} = \begin{cases} T_{mn->gw,old} & \text{intra-AD} \\ T_{mn->cn} + T_{cn->mn,old} & \text{cross-AD} \end{cases} \quad (18)$$

$$T_{rd,up} = 0 \quad (19)$$

Reverse tunneling is also possible in MIPv6. The end-to-end and redirection delays will then be the same as when using co-located COAs with reverse tunneling in MIPv4, see sections 3.2 and 3.5.

4. RESULTS AND CONCLUSION

To achieve low end-to-end and redirection delays, we are concerned about the number of times that a data or a signaling packet may have to traverse the backbone network. By use of the symbolic expressions presented in section 3, we can easily compare the different mobility support schemes. Except for the expressions concerning intra-AD handovers, each component in the expressions constitutes a potential backbone traversal.

- **End-to-end delay:** It is important to have $T_{ee,do}$ and $T_{ee,up}$ less than $T_{ee,max}$. As we have assumed that MN and CN use the same scheme, $T_{ee,do}$ and $T_{ee,up}$ will give the same number of traversals. From the equations in section 3 we can see that:
  - MIPv6 and MIPv4 with route optimization gives only *one* backbone traversal. (7,8)
  - Triangular routing can give *two* backbone traversals. (1,2)
  - Reverse tunneling can give *three* backbone traversals. (5,6)

If the network delays between the involved entities are large, the schemes using route optimization are the only ones that can achieve acceptable performance. Triangular routing and bidirectional tunneling may give long T_{ee} even in the case the MN and the CN are attached to the same subnet.

- **Downstream redirection delay:** The downstream redirection delay can differ if the handover is intra-AD or cross-AD:
  - For intra-AD handovers schemes based on hierarchical FAs or smooth handover are superior, since $T_{rd,do}$ does not include any backbone traversals in this case. (14,18)
  - For cross-AD handovers no scheme provides good support. MIPv4 with triangular routing, MIPv4 with reverse tunneling and MIPv6 (with route optimization or reverse tunneling) all risk losing packets corresponding to the delay of *two* backbone traversals. (3,18)

    For MIPv4 with route optimization the situation is even worse, since packets corresponding to *three* backbone traversals may be lost. (10)

    To improve $T_{rd,do}$ for MIPv4 with route optimization, the obvious approach would be to let the MN send a *BinUp* directly to the CNs, however, this is currently not suggested in [11].

    It should be noted that using hierarchical FAs is of no use for handovers across ADs, unless the hierarchy spans over several domains. Smooth handover mechanism are not likely to help either, since packets will have to traverse the backbone multiple times and possibly arrive to late at the MN.
- **Upstream redirection delay:** For the upstream redirection delay, we get different results if co-located or foreign agent COAs are used:
  - Schemes using co-located COAs have a $T_{rd,up}$ of zero. (16,19)
  - $T_{rd,up}$ for intra-AD handovers with hierarchical FAs will not involve any backbone traversals.(15)
  - Otherwise, schemes based on foreign agent COAs, the MN waits for a positive *BinAck*, leading to a $T_{rd,up}$ of *two* backbone traversals. (4,11)

It is worth noting that the MN receives information about an appropriate gateway on the new subnet in the *Agent Advertisements* sent out by the FA. Hence, it ought to be possible for the MN to send data (possibly simulcasting the packets to the old FA as well) to the CN via that gateway at the same time as it is able to send the *BinUp*, giving a $T_{rd,up}$ of approximately zero.

The overall conclusion is that MIPv6 is the scheme most well suited for services like IP based mobile telephony, since it gives the "lowest possible" end-to-end delay and relatively low redirection delay. Of the IPv4 schemes, MIPv4 with route optimization is the most promising one, due to its low end-to-end delay. To enhance it even further, the MN should be able to send *BinUps* to the CNs directly. Furthermore, to be able to cope with ingress filtering routers, support for reverse tunneling towards the CNs may be needed.

5. FUTURE WORK

In this study we have assumed that necessary keys for authentication etc has already been exchanged. However, to enable wide spread deployment of IP based mobility, a global key management infrastructure will be needed. This is particularly important for schemes where the MN sends *BinUps* directly to its CN(s). Related to this is the need for authentication, authorization and accounting (AAA) services[4]. To evaluate how this affects T_{ee} and T_{rd} is an interesting and important extension to this work.

The study could also be extended to give numerical results by measuring network delays for different distances and topologies and use these values in the equations in section 3. Furthermore, one should make measurements on (high performance) implementations of the mobility support schemes to study whether the associated processing delays are negligible or not.

Development of efficient link layer support will affect the results of this study. If the MN is able to connect to several APs simultaneously, the impact of T_{rd} will be reduced. T_{rd} will still be of importance, since it affects the need for cell overlap areas and fast movement detection mechanisms.

References

[1] J. Bound and C. Perkins. Dynamic Host Configuration Protocol for IPv6 (DHCPv6) <draft-ietf-dhc-dhcpv6-14.txt>, February 1999. Internet draft. Work in progress.

[2] R. Droms. RFC 2131: Dynamic host configuration protocol, March 1997.

[3] P. Ferguson and D. Senie. RFC 2267: Network ingress filtering: Defeating denial of service attacks which employ IP source address spoofing, January 1998.

[4] S. Glass, T. Hiller, S. Jacobs, and C. Perkins. Mobile IP Authentication, Authorization, and Accounting Requirements <draft-ietf-mobileip-aaa-reqs-03.txt>, March 2000. Internet draft. Work in progress.

[5] International Telecommunication Union Telecommunication Standardization Sector (ITU-T). Recommendation G.114, Transmission Systems and Media, General Characteristics of International Telephone Connections and International Telephone Circuits, One-Way Transmission Time, February 1996.

[6] G. Montenegro. RFC 2344: Reverse tunneling for mobile IP, May 1998.

[7] C. Perkins. Mobile-IP Local Registration with Hierarchical Foreign Agents <draft-perkins-mobileip-hierfa-00.txt>, February 1996. Internet draft. Work in progress.

[8] C. Perkins. RFC 2002: IP mobility support, October 1996.

[9] C. Perkins. RFC 2003: IP encapsulation within IP, October 1996.

[10] C. Perkins and D. Johnson. Mobility Support in IPv6 <draft-ietf-mobileip-ipv6-09.txt>, October 1999. Internet draft. Work in progress.

[11] C. Perkins and D. Johnson. Route optimization in mobile ip <draft-ietf-mobileip-optim-08.txt>, February 1999. Internet draft. Work in progress.

[12] S. Thomson and T. Narten. RFC 2462: IPv6 stateless address autoconfiguration, December 1998.

[13] J-O Vatn. Long Random Wait Times for Getting a Care-of Address are a Danger to Mobile Multimedia. In *1999 IEEE International Workshop on Mobile Multimedia Communications (MoMuC'99)*, November 1999. San Diego, CA, USA.

[14] J-O Vatn and G.Q. Maguire Jr. The effect of using co-located care-of addresses on macro handover latency. In *14th Nordic Teletraffic Seminar*, August 1998. Lyngby, Denmark.

Agent Based Seamless IP Multicast Receiver Handover

Jiang Wu and Gerald Q. Maguire Jr.
Department of Teleinformatics, Royal Institute of Technology (KTH), Sweden
{jiang, maguire}@it.kth.se

Key words: The fast development and deployment of the Internet and mobile

Abstract: The fast development and deployment of the Internet and mobile communication has boosted personal computing and communication. The ever-increasing number of personal devices accessing the Internet demands IP mobility support ubiquitously. However, the current TCP/IP protocol suite is not able to fully support IP mobility, especially when real-time applications are concerned. This paper discusses the IP mobility support in general and IP multicast receiver mobility support in particular. The current Internet is capable of providing basic mobility support for IP multicast when mobile hosts act as IP multicast receivers. However, the worst case handover latency introduced by IP layer handover is usually unacceptable to the real-time applications running in the mobile receivers. In order to overcome this, we propose a Mobility Support Agent (MSA) architecture and a set of protocols to help achieve seamless IP multicast mobility. Our testbed runs IGMPv2 and PIM-SM. Measurements have shown that by using the MSA architecture, the handover latency introduced by the current multicast membership management protocols and multicast routing protocols is nearly negligible.

1. INTRODUCTION

Today's mobile communication systems are primarily designed to provide cost effective wide area coverage for a rather limited number of users with moderate bandwidth demands (voice + low rate data). The users of tomorrow will expect much more than today's technology and infrastructure can offer. Research in [1] shows that pico-cells (very small radio cells) will be the dominant wireless access networks with high bandwidth capacity and low error rate. Both data applications and real-time applications will run over those pico-cells.

The new pico-cell based infrastructure will have IP running directly over the wireless networks. Figure 1 shows such an example network with link layer wireless networks connected via IP routers. With the help from IP, mobile communication can be maintained despite moves among heterogeneous wireless networks. Mobility support will not be limited to handover between the same link layer wireless network, rather, IP layer handover (vertical handover [2]) can hide changing of accessing network from applications as a mobile host roams.

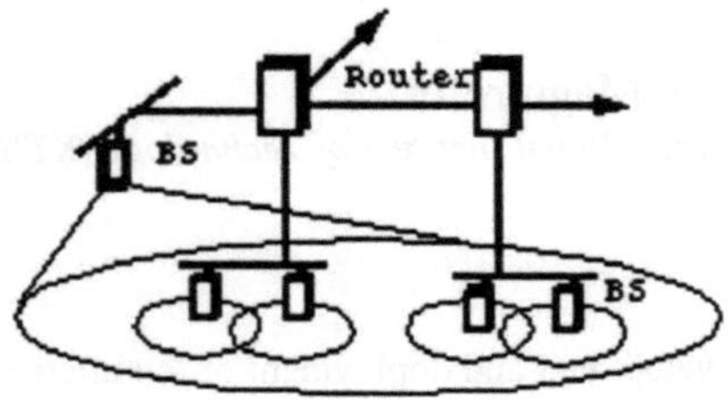

Figure 1. A pico-cell based wireless network infrastructure. Handover may occur in link layer or IP layer.

The key challenge for IP layer mobility support is from the TCP/IP protocol suite itself, due to the lack of handover support mechanisms. IP technology historically was developed and optimized for the fixed networks, though since its inception satellite networks were part of the network. Besides location dependent routing problems for IP unicast routing, Both IP unicast and IP multicast introduce significant amount of traffic loss during a handover (we call it "handover latency" in the following text), which makes it almost impossible to support real-time applications.

The unicast IP routing schemes use only the network part of the IP address to make routing decision for each packet, hence IP routing achieves very good scalability. However, those routing schemes implicitly assume that Internet hosts must attach to the network that has the same network prefix (the home network). Otherwise, it is impossible for the mobile hosts' home router to deliver the packets successfully to it.

The location dependent nature of unicast IP routing constrains the mobility of a host. Mobile IP [3] solves this problem by introducing an extension to the original unicast routing mechanism. When a mobile host visits a foreign network, the packets destined to that mobile host are intercepted by the Home Agent in the mobile host's home network. A tunnel is built between the Home Agent and a corresponding entity (either a

Foreign Agent in the foreign network or the mobile host itself) to deliver the packets to the mobile host. Mobile IP makes IP unicast handover possible. However, by using Mobile IP alone, it still introduces handover latency in the order of several seconds [4].

Unlike the unicast routing mechanism, IP multicast routing is location independent. An IP multicast address is a logical address representing a set of participating hosts in a multicast session. While in IP unicast, the address is used to indicate the point of attachment of a host to the Internet. A host simply propagates its interest in participating in a multicast session to the nearest upstream multicast router (not necessarily the one in its home network), in turn the intermediate multicast routers establish a new branch of the multicast tree which delivers the traffic to the host.

Handover latency for IP multicast varies from zero up to minutes. This paper examines the problem of inefficient IP multicast handover in the current Internet, and proposes an agent based architecture to help eliminate potential handover latency. A set of protocols were designed for this new architecture. Within our own testbed, simple implementation of the protocols and measurement of handover performance were done.

2. IP MULTICAST

IP multicast aims to provide a routing architecture to support group communication in a dynamic and efficient way. IP multicast consists of two parts: group membership management and multicast routing. The group management protocols work mainly in the leaf networks where the local membership information is collected, which is later used by the routing protocols to build up the multicast routing tree. The group management protocols should be able to quickly reflect the dynamic changing nature of membership in the leaf networks, while at the same time should consume as little as possible the network resources.

Different leaf networks usually have different group management mechanisms. One of the most popular protocols among them is the Internet Group Management Protocol version 2 (IGMP) [5], it is widely used in the LANs. Although other group management protocols are interesting, this paper focuses on IGMP. IGMP is used by the Querier Router to poll the presence of the multicast members in its directly attached LAN. The host may report its interest in a multicast group after receiving an IGMP Query from the Querier Router. The Querier Router consequently constructs a membership table indicating the presence of the members.

It is only important for the multicast routers to know whether a multicast group is active or not in the LAN (not the identities of the explicit hosts).

Thus the multicast group membership table in the router is not complicated. Report suppression is used to avoid report flooding when multiple participants report to the same group after an IGMP Query is received. Before sending an IGMP Report, a host backs off a random time interval. The host with the smallest timer will send out the report to that group, suppressing the other reports which are waiting for their timer expiration.

Multicast routing protocols can be generally divided into two categories: dense mode multicast protocols and sparse mode multicast protocols. As shown in Figure 2, the dense mode protocols and sparse mode protocols can be further divided into different groups. The resulting protocols: DVMRP [6], MOSPF [7], PIM-DM [11], PIM-SM [8] and CBT [9] are the ones currently getting most attention in the Internet. The inter-domain multicast protocol BGMP [10] is also under development nowadays.

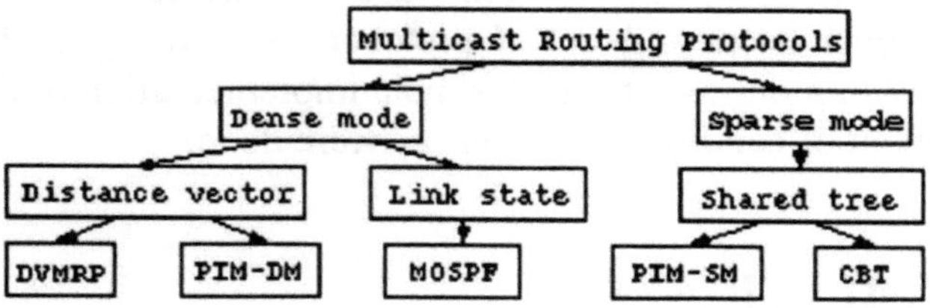

Figure 2. The multicast routing protocols

The dense mode and sparse mode routing protocols are different in many ways. Dense mode protocols are usually used in heavily populated member environments, while the sparse mode protocols are used in the situation when the group members are scattered over the Internet. Dense mode protocols build up the multicast tree usually by a data driven approach (MOSPF is an exception), while the sparse mode protocols requires an explicit join from the hosts.

When the first member joins a multicast group in the LAN, the Designated Routers will be informed by receiving an unsolicited IGMP Membership Report from that host. Immediately the Designated Routers will try to establish the multicast tree branch for the group. This is the mechanism used by IGMP and multicast routing protocols to reduce the joining latency for the first member.

The role of a multicast sender is quite different from the role of a multicast receiver. A host is not required to join in the multicast group when it acts only as a multicast sender. The host can directly send to the multicast group if there are multicast routers available in its directly attached LAN. Also the multicast routing protocols treat the sender differently from the receiver. This paper does not discuss sender mobility.

2.1 Handover for IP multicast receivers

Handover performance for IP multicast is determined by IGMP and multicast routing protocols. In this paper we use PIM-SM to illustrate the behavior of a routing protocol. Although PIM-SM is quite different from the other multicast routing protocols, they have similar behavior in handover.

Because IP multicast uses only the group address for membership reporting and routing, it doesn't require the hosts' identification in the network layer. The minimum requirement for a host to join a multicast group is that the host is attached to the same LAN as the corresponding Querier Router. A handover does not require new IP address as Mobile IP does. In principle, the current Internet is capable of handling IP multicast handover without any modification of the running protocols. However, the problem of handover for IP multicast comes from the potential heavy traffic loss. The handover latency in such a multicast handover is sometimes unacceptably long, i.e., 30 seconds.

After having established a steady multicast tree, IGMP and PIM-SM sends periodic protocol messages to maintain state in the host and multicast router. In IGMP, the Querier Router sends IGMP Query messages every IGMP Query Interval to probe the membership information in the LAN. In PIM-SM, routers send join/prune message upstream every Join/Prune-Period to refresh the routing states in the router. High granularity of the queries and join/prunes can make the protocols react quickly to the changing of the membership situation. However, the queries and join/prunes are sent only at a moderate frequency because otherwise it will lead to high protocol overhead. In order to respond quickly to a member's joining/leaving, both IGMP and PIM-SM introduce event triggered protocol messages.

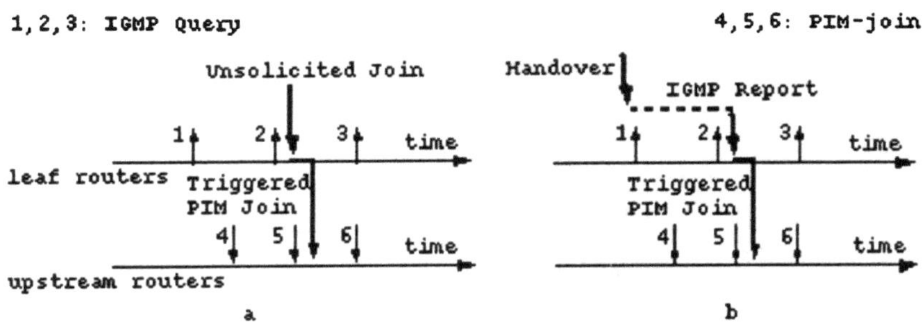

Figure 3. (a) An application first joins a multicast group. (b) An application joins after handover, in this case no member exists in the new IP network. A leaf router can be either Querier Router or Designated Router.

As can be seen in Figure 3a, When the application in the host wants to join a multicast group, an unsolicited IGMP Membership Report is sent. A routing entry for that multicast group, if not already existing, will be added in the PIM-SM Designated Router immediately. This new routing entry triggers the Designated Router to lookup the corresponding Rendezvous Point and send a join message to the upstream router. Thus when an application in the host first joins a multicast group, the joining delay is made very low by using the event triggered protocol messages.

After handover, a mobile host can continue to receive multicast traffic in the new IP network if there are other active members of the group in that network. However, it most probably will be the opposite case (the worst case) in a pico-cellar environment where the cell size is too small to accommodate many mobile hosts. Hence the probability of their already being a member of the multicast group in the cell is small. In order to receive the traffic quickly, a join message should be sent immediately regardless of the IGMP Queries. However, neither the multicast applications nor IGMP has the mechanism to detect the handover and trigger the unsolicited IGMP Membership Report.

Figure 3b shows that after a handover, the mobile host must wait for an IGMP Query in the new IP network. After receiving the IGMP Query, it sets up a random back off timer. An IGMP Membership Report will be sent when the timer expires.

The additional handover latency introduced by multicast routing protocols are not as severe as that due to IGMP, because the Designated Routers always use triggered join to the upstream router to establish the multicast tree branch. When a Designated Router discovers a new group is present in the LAN by IGMP, it will send immediately a join/prune message to the upstream router. It takes time to establish the tree branch. However, the tree establishment latency is low in most cases, i.e., when the distance between the Designated Router and the upstream router is short. In some cases, e.g., cross-ocean links, the propagation time to the upstream router is not negligible. But even in those cases it is less than the latency introduced by IGMP.

2.2 THE MSA ARCHITECTURE AND PRE-REGISTRATION

In this paper we propose a Mobility Support Agent (MSA) architecture to help achieve seamless IP mobility. The MSA architecture is designed to eliminate the potential traffic loss caused by handover. It is transparent to the applications running in the mobile hosts and the related protocols running in the network.

A set of protocols are designed for the MSA architecture. The MSA protocols run only between the MSAs and the mobile hosts. Although in this paper we only discuss the protocols related to IP multicast mobility, the MSA architecture can be used as general framework for IP mobility support. The architecture is shown in Figure 4a.

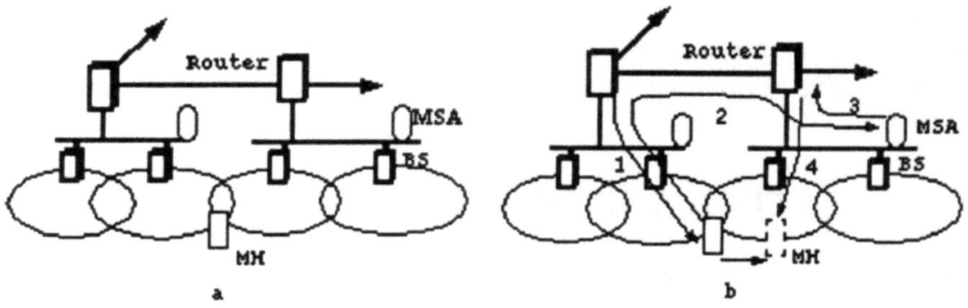

Figure 4. (a) The MSA architecture. (b) The procedure for the Pre-registration protocol. There is a MSA node located in each IP network which supports seamless IP mobility.

The key protocols for supporting IP multicast mobility are Agent Discovery Protocol and Pre-registration protocol. MSAs advertise their presence and services via the Agent Discovery protocol. The MSAs form their own multicast group to which they are both senders and receivers. Periodic protocol message exchanges update the MSAs' current information about each other. The advertised information may include the MSAs' IP addresses, the services they provide, etc. The advertisement between the MSAs and mobile hosts is exchanged locally via ICMP Router Advertisement Extensions.

The Pre-registration protocol is designed for supporting seamless IP multicast. A Pre-registration procedure is illustrated in Figure 4b.

1. A mobile host receives multicast traffic from the Internet in its currently attached network.
2. As soon as the mobile host decides to perform a handover, it sends a Pre-registration message using UDP to the neighboring network's MSA. This Pre-registration packet is routed according to the unicast routing protocols.
3. Enclosed in this packet are the multicast groups the mobile host is participating in. Once having received this packet, the MSA sends immediately an IGMP Membership Report to the Designated Router, which in turn triggers a join message to the upstream routers to establish the multicast tree.
4. When the mobile host arrives at the new network, multicast traffic is already available for it.

One advantage IP multicast has over IP unicast in handover is that the mobile host knows in advance the multicast addresses it will use in the new network are those same addresses as it used before the handover. The multicast traffic grafting is from the closest upstream router, while in Mobile IP, traffic has to be redirected by the fixed home agent.

The Pre-registration protocol can be extended to support seamless Mobile IP support. The additional functions that should be added in the MSA are: pre-negotiate the co-located care-of address and pre-register with the home agent for traffic redirecting, etc.

3. TESTBED AND MEASUREMENT

Figure 5 shows the testbed we use for testing the MSA architecture. The Pre-registration protocol is implemented in the testbed. Routers 1, 2 and 3 all support PIM-SM and all the nodes in the testbed support IGMPv2. In the testbed (as shown in 5a), A multicast sender “Msender” sends packets at a constant rate to a certain multicast group. The mobile host first joins the multicast group in network 2. At a certain moment the mobile multicast receiver "MH" handovers to network 3. Our measurement will compare the worst case handover performance with/without using the MSA in network 3.

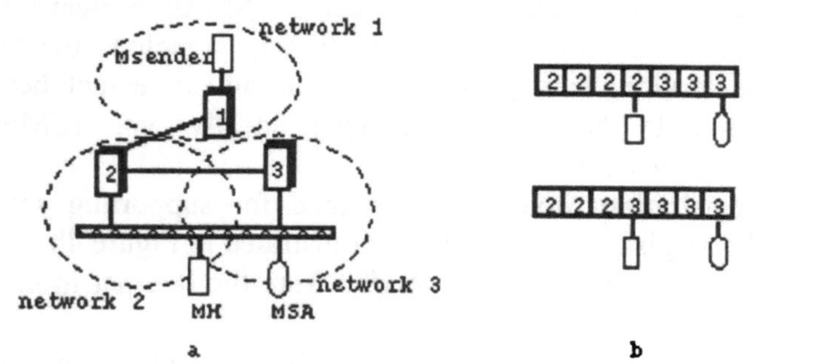

Figure 5. (a) The MSA testbed. (b) Handover emulation using Ethernet.

We use 10BaseT Ethernet in our testbed, each Ethernet segment is a separate IP network. The Ethernet segments emulate pico-cells. Handover is emulated by moving the mobile host from one Ethernet segment to another. The HP hub we use has a control module that can dynamically allocate the port to different Ethernet segments. As shown in Figure 5b, the port the mobile host is attaching to can be switched from network 2 to network 3 by using a SNMP control message.

3.1 Handover latency in handovers without pre-registration

Figure 6 illustrates the handover procedure for IP multicast, when there is no members in the new netwrok. The handover latency (T_ho) consists of physical movement time when non-overlapped cells are used (T_mv), IGMP Query waiting time (T_qw), IGMP Report back off time (T_rb) and multicast tree establishing time (T_te).

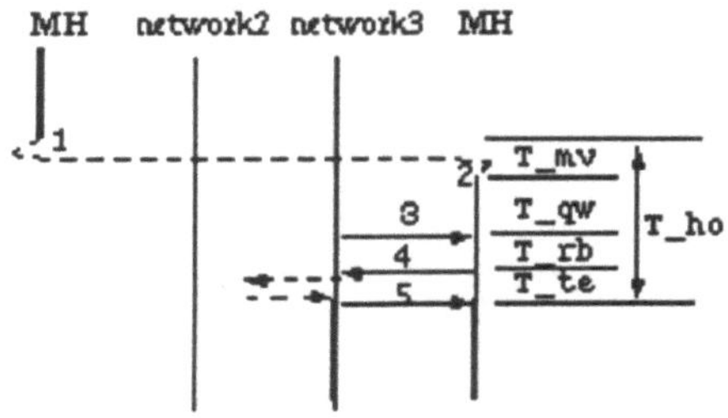

Figure 6. Mobile host handovers in normal IP multicast. The dash arrows between 4 and 5 indicate the tree establishment process. Thick line indicates multicast traffic.

The corresponding events during a handover are:
1. The mobile host leaves network 2.
2. The mobile host enters network 3.
3. The first IGMP Query is received from the Querier Router.
4. An IGMP Membership Report is sent after the back off timer expires.
5. The first multicast packet received in network 3.

In our test, the mobile host was in a stable situation receiving the multicast packets before the handover. Because handover is emulated by switching the Ethernet hub port by using SNMP, T_mv is very small (around several milliseconds) so that it can be neglected. Our focus will be on T_qw, T_rb and T_te.

In this measurement , the Mender sends 50-Byte packets once per second. Because the handover latency in a normal handover is in the order of seconds, this packet rate is high enough to capture the handover latency characteristics. "Tcpdump" is used in network 3 to capture the events during a handover.

Router 2 and router 3 send all IGMP queries every 60 seconds (this is done using Cisco 7000 series routers). The mobile host may handover at any moment during the IGMP Query Interval, so the waiting time for the next IGMP Query can be anywhere between 0 and 60 seconds. When the mobile host receives the query, it sets a random timer between 0 and 10 seconds, for sending IGMP Membership Report.

Figure 7 shows the measurement results. We observe that T_qw and T_rb contribute most to the overall handover latency. The maximal T_qw in our test is 51 seconds, and the average is around 27.2 seconds. The average T_rb is around 5.2 seconds. Due to IGMP, a mobile host needs to wait for 32.4 seconds on average to continue to receive the multicast traffic after a handover.

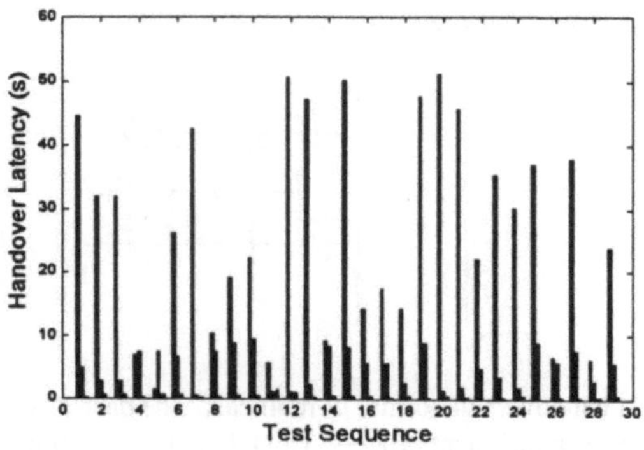

Figure 7. Measurement results of handover latency. The bars represent T_qw, T_rb and T_te from the left to right order for each group data, T_mv is negligible and not shown in this figure.

T_te is considerably smaller when compared to T_qw and T_rb, with an average of 0.5 second. In the result T_te is calculated by comparing the time difference between step 5 and step 4. The real T_te could be even smaller if the Msender sends out packets at higher rate than in these measurements. On the other hand, the small T_te results form the short distance between the upstream router (router 2) and the leaf network (network 3). It will be larger when the upstream router is far away or the network is highly loaded. However, we expect that T_te will not have the same significance in overall IP multicast handover latency.

We can see from the measurements that because of the large handover latency during an IP multicast handover, the current Internet simply can not support any of the applications that require real-time communication.

3.2 Handover latency in handovers with pre-registration

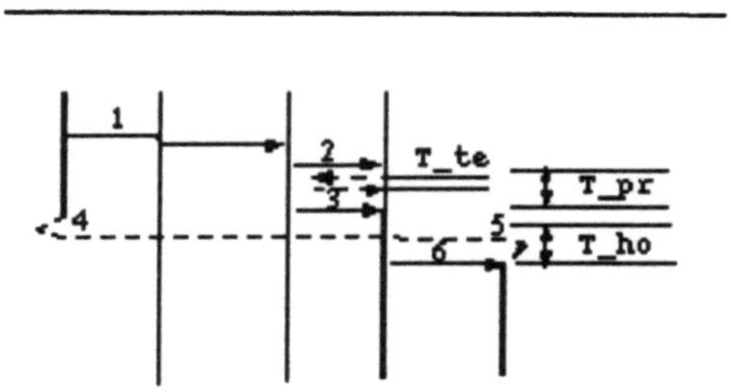

Figure 8. Mobile host handovers with pre-registration.

Pre-registration eliminates the need for waiting for T_qw and T_rb by asking the MSA to send an unsolicited IGMP Membership Report in advance of the handover. In consequence, T_te is also made to happen before the mobile host moves to the new network. The detailed handover procedure with pre-registration is illustrated in Figure 8:

1. The mobile host sends a pre-registration message to the MSA in network 3.
2. The MSA joins the multicast groups as a proxy for the mobile host.
3. The first multicast packet arrives in network 2.
4. The mobile host leaves network 2.
5. The mobile host enters network 3.
6. The mobile host starts to receive multicast traffic in network 3.

As can be seen, multicast traffic arrives at network 3 before the arrival of the mobile host. Thus the latency the mobile host suffers from mainly comes from the physical movement between the non-overlapped networks and other possible disturbances by the handover, e.g., network overload caused by additional multicast traffic. In some cases when the upstream router in the new network is closer to the Rendezvous Point, receiving duplicated packets is possible, but this is not a big problem for IP multicast receiving.

Determining the handover latency in the measurements, by sending a packet once per second and using "Tcpdump" in network 3 is not feasible. Hence, we conducted a second test, in this test the Msender sends 50-Byte packets every millisecond, producing a 400kbps stream traffic. The Msender puts a sequence number in each packet when sending. The mobile host receiving these packets will record the sequence number and calculate how many packets are missed during the handover, the packet loss corresponds to the T_ho in Figure 8. In this way we can measure the handover latency with resolution of 1 millisecond.

Measurement results are shown in Figure 9. From these results we observe that 34 percent of the handovers suffer no traffic loss. The handover latency is up to 11 milliseconds, the average is 5.1 milliseconds. This result is significantly less than the result in the previous measurements (32.4 seconds) when no MSAs were used. We believe that by using the MSA architecture and Pre-registration, real-time applications can operate despite making a handover.

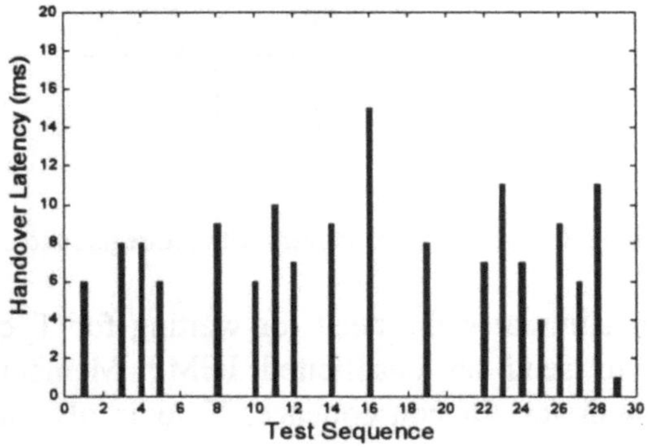

Figure 9. Measurement results of the handover latency with pre-registration.

In this scenario, physical movement is the major factor contributing to the overall handover latency. In our testbed, even though the "cells" are not overlapped, the physical movement time is almost negligible because of the mechanism we use. In other cases when the distance between the cells are large, the physical movement will cause noticeable disturbance to the applications. However, good network planning can solve this problem. Another possible solution is to add features to the Pre-registration protocol to ask the MSA to buffer the incoming multicast packets which can be later delivered to the mobile host after it has handedover to the new network.

4. CONCLUSIONS AND FUTURE WORK

Expecting a pico-cell based high capacity wireless infrastructure in the future, we find that handover will frequently happen at the IP layer. By introducing the MSA (Mobility Support Agent) architecture to the existing Internet routing architecture, seamless mobility support can be achieved. In this paper, we show that no additional processing or elements are needed for supporting mobility for multicast receivers. However, in order to make low latency handover, MSAs are used. Tests and measurements have shown that the worst case handover latency drops to a level that is sufficient to support even quite demanding real-time communication during a handover.

We are going to explore the handover procedure in more detail, better implementation and more measurement will be done in the near future. Real-time video will be used in our testbed to test the feasibility of mobility support for the applications.

In order to provide better performance and to be flexible in pre-registration, movement detection and prediction need to be studied more. How to reduce the departure latency in the previously visited network is also an interesting problem. Finally, the handover behavior as a multicast sender will be studied to make the multicast mobility research complete.

REFERENCES

[1] M. Flament, F Gessler, F Lagergren, O Queseth, R Stridh, M Unbehaun, J Wu, J Zander, "An Approach to 4th Generation Wireless Infrastructures - Scenarios and Key Research Issues", VTC 99, May 1999.

[2] Mark Stemm, Randy H.Katz, "Vertical handoffs in wireless overlay networks", ACM Mobile Networks and Applications", Volume. 3, 1998, pp. 335-350.

[3] C. Perkins, "IP mobility support", RFC2002, October 1996.

[4] Jon-Olov Vatn and Gerald Q. Maguire Jr. "The effect of using co-located care-of addresses on macro handover latency", Fourteenth Nordic Tele-traffic Seminar, August 1998.

[5] W. Fenner, "Internet group management protocol, version 2", RFC2236, November 1997.

[6] D. Waitzman, C. Partridge, S. Deering, "Distance vector multicast routing protocol", RFC1075, November 1988.

[7] J. Moy, "Multicast Extensions to OSPF", RFC1584, March 1994.

[8] D. Estrin, D. Farinacci, A. Helmy, D. Thaler, S. Deering, M. Handley, V. Jacobson, C. Liu, P. Sharma and L. Wei, "Protocol independent multicast-sparse mode (PIM-SM): protocol specification", RFC2362, June 1998.

[9] A. Ballardie, "Core based trees multicast routing, protocol specification", RFC2189, September 1997.

[10] D. Thaler, U. Michigan, D. Estrin, D. Meyer, U. Oregon, "Border Gateway Multicast Protocol (BGMP): Protocol Specification", Internet Draft, March 2000.

[11] S. Deering, D. Estrin, D. Farinacci, V. Jacobson, A. Helmy, and L. Wei, "Protocol independent multicast version 2, dense mode specification", Internet Draft, May 1997.

Predistortion for Solid State Amplifier of Mobile Radio Systems

Henryk Gierszal, Witold Hołubowicz and Przemysław Sułek
Institute of Communication and Information Technologies in Poznań,
Palacza 91A, 60-273 Poznań, Poland
University of Technology and Agriculture in Bydgoszcz
Kaliskiego 7, 85-763 Bydgoszcz, Poland

Key words: predistortion, nonlinearity, mobile system, amplifier, compensation for nonlinearity, 3rd generation system, UMTS, GSM, TETRA, TWT, SSP, satellite

Abstract: In this paper, we implement memoryless data predistortion in QAM trasmission over Gaussian channel. The main advantage of predistortion is the ability to reduce the back-off of an amplifier without loss of performance. Comparing three types of Solid State Amplifier, one of them distinguishes significantly better performance taking into account global degradation and bit error rate.

1. INTRODUCTION

The present world of telecommunication is transmitting an increasing amount of information in digital form. The number of services offered grows rapidly. Thus, higher rates of transmission and simultaneously higher quality are becoming standard requirements for different modern systems, such as pulse coded modulation for multiple voice telephony, digital audio broadcasting or digital high definition television, and 2nd as well as 3rd generation mobile radio systems like GSM, TETRA or UMTS. In order to reach high rates, there is, however, a second aspect related to the growing number of services that telecommunications must cope with. The electromagnetic frequency band, as a resource, is unique and must be used wisely, e.g., efficiently.

The modern advanced techniques like CPM, TCM or Turbo-Codes do not solve all problems which appear when the number of states in signal constellations increases. As the number of levels grows, modulation is more sensitive to propagation conditions like fading, echoes and noise as well as other linear and non-linear distortion. Among different system imperfections, there are:

- carrier frequency and phase errors due to the phase jitter of a phase-lock loop [1] at the receiver,
- errors due to a jitter of the optimum sampling instant,
- sinusoidal interference generated by a local oscillator whose frequency falls within the received signal bandwidth,
- filtering imperfections which produce linear and sinusoidal distortions depending on the amplitude and phase response of the channel filters,
- non-linear distortion.

Non-linear distortion is primarily due to the transmitter high-power amplifier (HPA) which limits considerably the bandwidth-efficiency and performance. The effect of HPA nonlinearity can be reduced by backing-off the output signal level from the amplifier's saturation point, but it reduces the transmitted signal power and, consequently, the link flat-fade margin. The link margin is a safety margin introduced in the system to improve efficiency against random disturbances. To increase this margin, the HPA should be driven as close to its saturation point as possible. Therefore, the transmitted signal power is chosen, in practice, as a trade-off between these conflicting two requirements, i.e., high amplification and low signal distortion.

The basic purpose of this work is to examine a predistortion technique for different types of HPA. Studies were done for non adaptive memoryless data predistortion. All methods are tested for 16-, 32- and 64-QAM signal constellations. The roll-off factor in the raised-cosine transmit and receive filters takes value of 0.35.

2. CHANNEL DESCRIPTION

2.1 Radio System

The radio system operating over an AWGN channel is considered. The channel has no multipath propagation, i.e., the transmission medium only adds white Gaussian noise n(t) to the transmitted signal of symbols $\{a_n\}$. Furthermore, the timing and carrier synchronization circuits are assumed ideal, so the radio system can be represented by a simple equivalent baseband model. Omitting any predistortion at the transmitter or any compensation technique at the receiver, a block diagram of the radio system is sketched in Fig. 1(a), and its baseband-equivalent in Fig. 1(b), where T is the symbol interval and $\delta(\cdot)$ is the Dirac delta function.

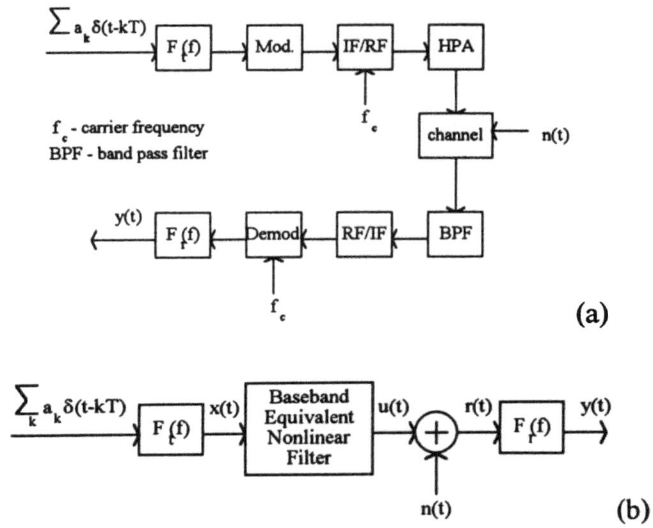

Fig. 1. (a) Block diagram of the considered system. (b) Its baseband-equivalent model.

2.2 Radio System Modules

2.2.1 Filtering and Modulations

The pulse shaping is performed at baseband and is assumed to have a raised-cosine Nyquist characteristic evenly split between the transmitter (filter $F_t(f)$) and the receiver (filter $F_t(f)$). In the absence of HPA nonlinearity, data transmission is free of intersymbol interference ISI, and the signal-to-noise ratio SNR is maximised at the sampling instants (matched filtering). This splitting minimizes also the overlap between adjacent channels.

In Fig. 1(a), a band pass filters BPF is sketched which cancels unwanted products coming from adjacent channels. A BPF can be placed after the HPA to compensate for the spectral spreading due to the HPA itself but it introduces an additional loss of HPA power.

Successive signal states of QAM modulations are uncorrelated, and maximum-likelihood sequence estimation (MLSE) reduces to symbol-by-symbol threshold detection.

2.2.2 Amplifiers

Two types of HPA's commonly used in microwave systems are travelling-wave tube (TWT) HPA' s and GaAs FET HPA' s (called as Solid State Power SSP). It is known that TWT HPA's present even less linear characteristics than SSP amplifiers. The HPA is assumed to have a frequency-independent memoryless bandpass nonlinearity characteristic [2]. The TWT nonlinearity is completely characterised by its AM/AM conversion

$$a(t)=\frac{\alpha_A r(t)}{1+\beta_A r^2(t)} \tag{1}$$

and AM/PM conversion

$$\psi(t)=\phi(t)+\frac{\alpha_P r(t)}{1+\beta_P r^2(t)} \tag{2}$$

where r(t) is the magnitude and $\phi(t)$ is the phase of a point in the input constellation. α_A, α_P, β_A and β_P are coefficients identified different TWTs. Output magnitude a(t) and phase $\psi(t)$ depend only on the current value of input magnitude r(t).

The SSP amplifier is characterized by the following equations:

$$a(t)=\frac{v_k r(t)}{\left[1+\left(\frac{v_k r(t)}{A_0}\right)^{2p_k}\right]^{\frac{1}{2p_k}}} \tag{3}$$

and

$$\psi(t)=\phi(t) \tag{4}$$

The AM/PM transition is linear while the output magnitude a(t) is function of an input magnitude and some coefficients changing with the type of SSP.

To quantify the signal degradation introduced by backing-off the amplifier from its saturation point, the output back-off factor, BO_o, is defined by the ratio of the transmitted average signal power, P_o, to the HPA's output saturation power, P_{sat}, e.g.,

$$BO_o\big|_{dB}=10\log\frac{P_{sat}}{P_o} \tag{5}$$

The study has focused on three SSP amplifiers. Two SSP amplifiers are by Sony and the third one (RF2108) is by Micro Devices. Figures 2 ÷ 4 show the amplitude and phase characteristics given by producers' catalogs.

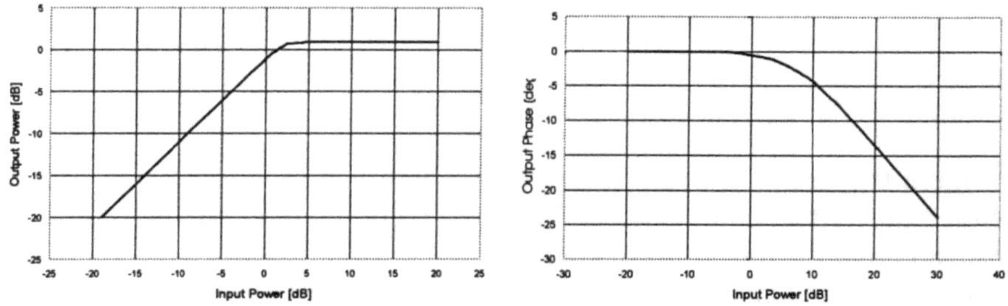

Fig. 2. Amplitude and phase characteristics of Sony's SSP1 amplifier.

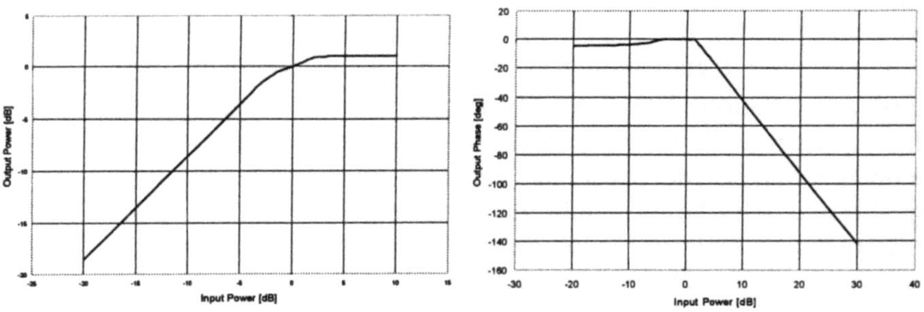

Fig. 3. Amplitude and phase characteristics of Sony's SSP2 amplifier.

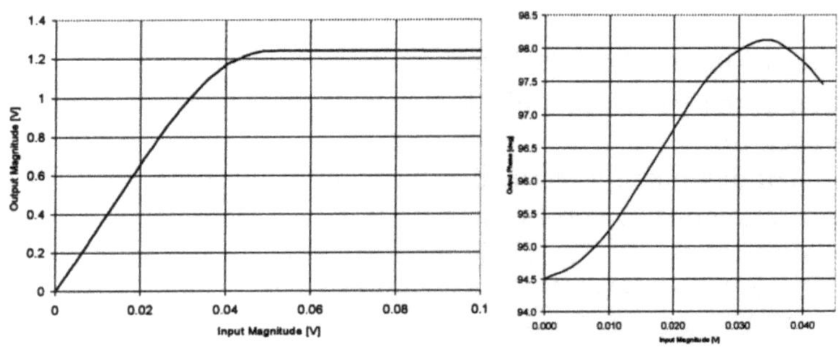

Fig. 4. Amplitude and phase characteristics of RF2108 amplifier.

2.3 Channel Model

Although the HPA is modelled with memoryless nonlinearity, its embedment between linear filters leads to a non-linear system with memory as a model for the overall channel. A convenient tool to represent such a system is the Volterra series [3].

The receiver filter output y(t) (as shown in Fig. 1) can be expressed as

$$y(t)=\sum_{k=0}^{\infty}\left[\frac{\binom{2k+1}{k}}{2^{2k+1}}\right]\sum_{n_1}\sum_{n_2}\cdots\sum_{n_{2k+1}}a_{n_1}^{*}a_{n_2}^{*}\cdots a_{n_k}^{*}a_{n_{k+1}}\cdots a_{n_{2k+1}}\cdot \tag{6}$$

$$\cdot\, h_{2k+1}\left(t-n_1T,t-n_2T,\ldots,t-n_{2k+1}T\right)+N(t)$$

where the terms in brackets represent the binomial coefficients, the a_n's are the transmitted data symbols, T is the symbol interval, N(t) is an additive noise, and h_{2k+1} is the Volterra coefficient.

In QAM systems, the nonlinearity has two major effects. First, the clustering phenomenon reflects the linear and non-linear ISI. We obtain a cluster of points instead of a single point in the constellation. In a non-linear system, the filtering is responsible for this effect. The second effect of the nonlinearity is that the respective gravity centers of the various clusters are no longer on a rectangular grid. This distortion is called warping of the signal constellation. These two phenomena depend on the roll-off factor α, which influence the memory of the system. As α decreases, the clustering becomes stronger. Moreover, the size of the clusters tends to be larger for signal points with larger amplitude. This confirms the observation from the Volterra series analysis [4] indicating that the ISI energy is concentrated in products involving the present symbol.

3. PERFORMANCE ANALYZE

For each modulation scheme, the carrier-to-noise ratio (CNR) degradation at the BER value of 0^{-3} was computed for different values of the HPA back-off.

The global degradation, defined as the sum of the HPA back-off and the CNR degradation due to non-linear distortion was then plotted as a function of the amplifier back-off. The CNR degradation is calculated in comparision with the channel with AWGN. The resulting set of convex curves gives the optimum HPA back-off for each system as well as the corresponding degradation. This measure is related to the flat-fade margin of the radio link. For an accepted degradation level one can choose a system which offers the smallest back-off.

The computation of the CNR degradation was based on generating a pseudo random data sequence, computing the corresponding discrete channel output sequence, and estimating the BER for specified CNR values. BER estimation was made using a quasi-analytic method [5], which estimates the error probability as mean of conditional probabilities that a received symbol without noise falls in an area determined by the treshold detector for the desired symbol. The quasi-analytic technique requires less symbols to evaluate the BER and is more precise than Monte-Carlo method.

Constellations for each modulation scheme are observed at the receiver. The same condition as for the computation of global degradation, e.g., the same back-off, must be present to compare the degradation of each constellation.

The evaluation of the BER as a function of the SNR was performed using the optimal back-off read from the global degradation curve. Additionally, the influence of HPA back-off degradation was taken into consideration, e.g., the x-axis represents the sum of the SNR and the amplifier back-off, because the back-off can also be interpreted as SNR degradation.

4. PREDISTORTER

To make better use of the available HPA power, some compensation techniques can be used at the transmitter or at the receiver. Normally it is better to place compensation circuits at the transmitter because it deals with a noiseless signal. The use of fixed or adaptive analog signal predistortion at IF or RF stage of the transmitter has almost become common practice in digital microwave radio systems. Adaptive predistortion has the advantage of automatically coping with any time-variations of the HPA response due to temperature variations and aging, while the intervention of a human operator is occasionally necessary in the case of fixed predistortion. Furthermore, predistortion has to be adaptive if it needs to be incorporated in a transmitter with automatic control of the transmitted power like in base stations and in mobile units of almost all mobile systems. Another approach to handle the non-linear distortion problem is the data predistortion which consists of modifying the input signal constellation so as to obtain the desired constellation after the HPA nonlinearity.

4.1 TWT predistorter

The key idea is that if the HPA nonlinearity can be characterized by a pair of functions as given by (1) and (2), the desired signal constellation can be obtained at the (demodulated) HPA output by driving the modulator by an appropriately predistorted constellation.

By computing a pair of (predistorted) coordinates for each point of the desired constellation, the effect of HPA nonlinearity can be eliminated completely. This is true when all pulse shaping is performed after HPA, but in this case, where some filtering precedes the HPA, data predistortion can only compensate for constellation warping. It does not reduce (and experience has shown that it can even enhance) the clustering of the signal constellation which reflects non-linear ISI. The baseband block diagram of this technique is depicted in Fig. 5.

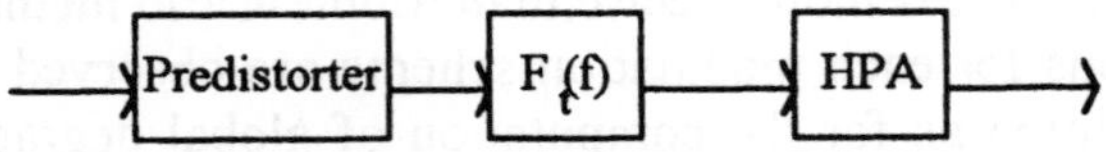

Fig. 5. Baseband block diagram of memoryless data predistortion.

The inverse of the TWT equations (1) and (2) leads to the following ones

$$r(t)=\frac{\alpha_A}{2c(t)\beta_A}\left[1-\sqrt{1-\left(2c(t)\sqrt{\beta_A}\right)^2}\right] \text{ if } c(t)<\frac{\alpha_A}{2\sqrt{\beta_A}} \quad (7)$$

and

$$\phi(t)=\varphi(t)-\frac{\alpha_P c^2(t)}{1+\beta_P c^2(t)} \text{ if } c(t)<\frac{\alpha_A}{2\sqrt{\beta_A}} \quad (8)$$

4.2 SSP predistorter

The predistorter characteristic is an inverse function of the SSP amplifier equations (3) and (4).

$$r(t)=\frac{1}{v_k}\frac{c(t)}{\left[1-\left(\frac{c(t)}{A_0}\right)^{2p_k}\right]^{\frac{1}{2p_k}}} \quad \text{when } c(t)<A_0 \quad (9)$$

$$\phi(t)=\varphi(t) \quad (10)$$

The Tab. 1 contains the values of parameters of the predistorter obtained with least-square fitting of equation (3) to real characteristic (figures 2 ÷ 4) of SSP amplifiers.

Table 1. Parameters of predistorters for SSPs.

Amplifier	v_k	A_0	p_k
SSP1	1.65	1	3.4
SSP2	1.82	1.04	1.65
RF2108	1.3	1	5

For simulation purposes, all characteristics of amplifiers and predistorters were normalized in order that the saturation point has the coordination of (1,1).

5. SIMULATION RESULTS

Using QAM signal formats, we made computer simulations to compare SSP amplifiers. In what follows, we give the main results of a simulation study which aimes at comparing predistortion technique on Gaussian channel. The simulations were carried out using a QAM system with ZF equalisation.

Figures 6 and 7 show global degradation for 16- and 64-QAM and three types of discussed amplifiers. The performance of an applied predistorter is presented on figures 8 and 9.

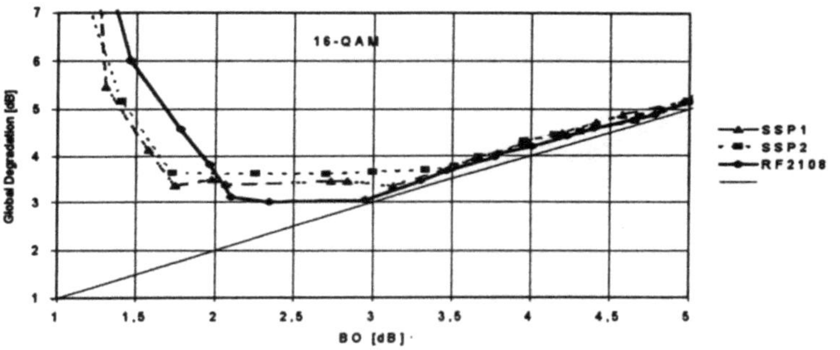

Fig. 6. Global degradation for 64-QAM modulation scheme and three types of amplifiers. Predistorter is on.

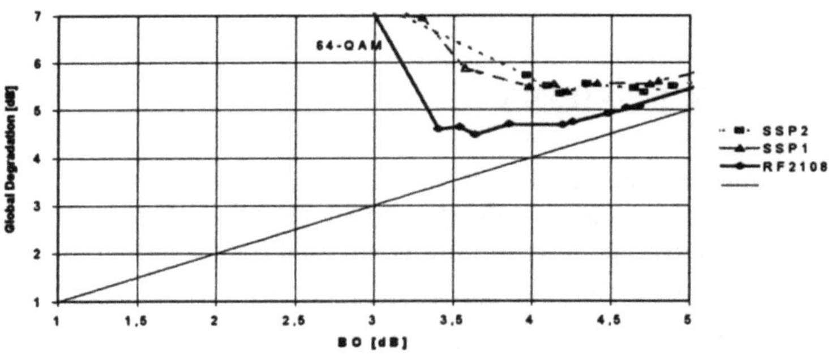

Fig. 7. Global degradation for 64-QAM modulation scheme and three types of amplifiers. Predistorter is on.

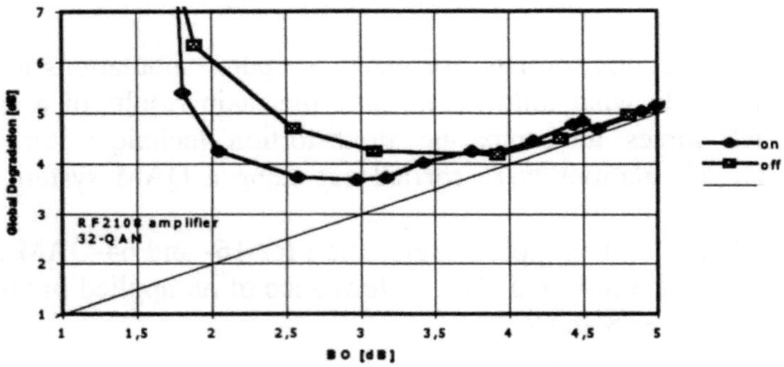

Fig. 8. Comparison of global degradation when predistorter is on and off for 32-QAM.

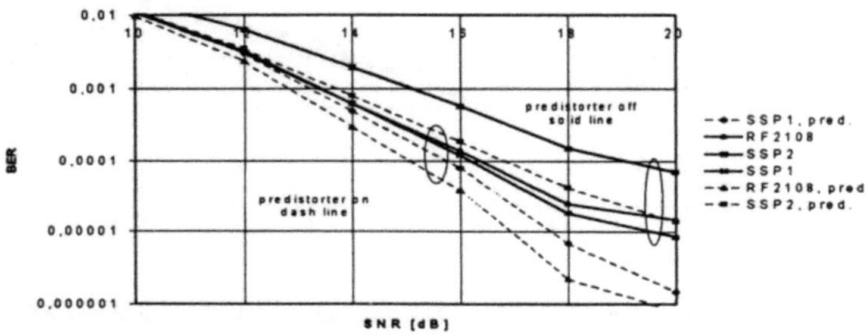

Fig. 9. Comparison of BER when predistorter is on and off for 32-QAM.

6. CONCLUSIONS

The memoryless data predistortion is easy to implement and offers some gain compared to a system without predistortion but it cannot compensate for the memory introduced by the filters situated before and after the HPA.

Using that predistorter, the global degradation can be reduced by up-to 1 dB and back-off may be less of about 1.5 dB. This yields better BER performance. For BER equal 10^{-4} using 32-QAM modulation scheme one can gain about 2.5 dB for SSP2, 1.3 dB for RF2108 and only 0.2 dB for SSP1 amplifier.

Among considered Solid State Amplifiers, one of Micro Devices offers the best performance, e.g., less back-off and less degradation than the amplifiers by Sony.

That modern technique is very useful to help us achieve the goal of transmitting the largest amount of information possible on the available bandwidth. The new bandwidth efficient techniques can be used even more efficiently on systems such as mobile radio, or other hertzien type channels for which the price of the bandwidth and number of users keeps increasing.

REFERENCES

[1] J. G. Proakis, "Digital Communications," New York: McGraw-Hill, 1983.
[2] G. Karam and H. Sari, "Generalized Data Predistortion Using Intersymbol Interpolation," Philips Journal of Research, vol. 46, no. 1, pp. 1-22, 1991.
[3] R. D. Gitlin, J. F. Hayes and S. B. Weinstein, "Data Communications Principles," New York: Plenul Press, 1992.
[4] G. Karam, "Analyse et compensation des distortions non linéaires dans les faisceaux hertziens numerique", Télécom Paris (ENST), octobre 1989.
[5] M. J. Jeruchim, "Techniques for Estimating the Bit Error Rate in the Simulation of Digital Communication Systems," IEEE JSAC, vol. SAC-2, pp. 153-170, January 1984.

Adaptive Antenna Technique for Mobile Communication

Ryszard J. Katulski
Technical University of Gdansk, Department of Radiocommunication
e-mail: rjkat@sunrise.pg.gda.pl

Keywords: mobile telecommunication, adaptive antenna system, direction of arrival procedures, microstrip technology.

Abstract: In this paper the adaptive array properties analysis due to its application in a base station of cellular mobile telecommunication system is presented. First, taking into account the interference problem, the base station antenna system with adaptive directional properties in horizontal plane is motivated. The functional structure of the adaptive processor is described. Next, the problem of the direction of radio signal arrival (doa) estimation have been presented. Especially, the MUSIC and ESPRIT doa algorithms are analysed. A structure of the adaptive array in microstrip technology is proposed. In conclusion, the usefulness of the adaptive technique in a base station of mobile cellular telecommunication is confirmed.

1. INTRODUCTION

The land mobile communication, especially cellular telephone networks are very modern form of telecommunication systems. Mobile telecommunication in general and cellular service in particular have been forefront of research and development activities. The progressive development of the wireless telecommunication systems, which are realized by use the radio link equipments, connects with the increaising of an interference level. The limitation of frequency band and the interference problem are important questions connected with cellular mobile systems, particularly in the future high capacity systems. A solution of the problems can be obtain by application the frequency reuse and cell splitting as well as cell sectorization with directional form of an antenna radiation pattern. These require to use for base station of the mobile network special antenna system, especially an adaptive antenna array technique with single narrow beams dynamically assign to iluminate the mobile units [1].

The researches and development activities connected with mobile communication systems, including the adaptive antenna theory and technique, have been done in Radiocommunication Department of the Technical University of Gdansk, Poland [2].

2. INTERFERENCE PROBLEM IN MOBILE RADIO SYSTEM

The main element of each mobile wireless system is the radio link consists of a base station and mobile unit, between which a radiowave propagation medium is existed. The antenna equipments installed in both base and mobile unit are interfaces between the wire and wireless part of the radio link, which structure in a short functionally form is presented in Figure 1.

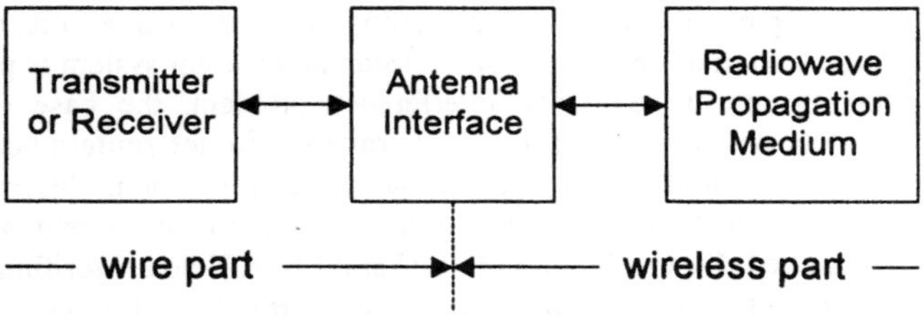

Figure 1. Functionally form of a radio link structure

Properties of these antennas influence on the radio link operation condition, especially the shape of these antennas radiation patterns, decide about the values of the interference signals. Each transmitter equipment is a source of the interference signals, which space distribution depends on the transmitter antenna directivity properties. Moreover, the directivity properties of an antenna operated with a receiver equipment influence on the received interference signals. Therefore, the interference signal level depends on a shape of the antenna radiation pattern [3].

Taking into account the movements of the mobile units the simple antenna form with omnidirectional radiation pattern is prefered to apply in a base station and mobile units. But, the directivity properties of the simple antenna structure are not optimal, especially for the base station of the high capacity cellular networks. The intereference problem requires to use in the base station the directional antenna with multi-beam radiation pattern for multichannels operation, in which each beam iluminates the individual mobile unit. This requirement can be realized by adaptive antenna system with mobile beams. For example, if the probability ratio $p_{one-beam}(s_w \leq s_i + p_r)$ at mobile unit, for omnidirectional radiation pattern in a base station, is assumed as:

$$p_{one-beam}(s_w \leq s_i + p_r) = p, \tag{1}$$

where: s_w - value of the wanted signal,
s_i - value of the co-channel interference signal,
p_r - protection ratio,

the probability ratio $p_{multi-beam}(s_w \leq s_i + p_r)$ at mobile unit, for mutli-beam radiation pattern in base station, is resulted as:

$$p_{one-beam}(s_w \leq s_i + p_r) = \frac{p}{m}, \tag{2}$$

where m is number of beams.

In result, the directional antennas are convenient to use in the base stations of a cellular telecommunication network, especially with multi-beam radiation pattern for multi-channels operation in order to obtain the better value of the spectrum utilization efficiency, i.e. number of channels per megahertz per square kilometer [4].

3. ADAPTIVE PROCESSOR

An adaptive antenna system is the antenna array that controls its own radiation pattern, by means of feedback control loop, while an adaptive processor operates. Adaptive antenna systems have been employed in military technique. Application in civil communication is not widespread, although some potential schemes have been suggested [5, 6].

A short form of the adaptive processor system is presented in Fig. 2, here the set $\{x$ of radio signals received by each antenna elements is multiplicated by set $\{w\}$ of complex weights, and next are summed to produce an output signal. The output signal y is compared with reference signal r and in result the error signal e is obtained to steering with set $\{x$ the processor unit [7].

The dynamic form of the output signal can be presented in matrix notice as:

$$y(t) = \mathbf{w}^{\mathrm{T}} \cdot \mathbf{x}, \tag{3}$$

where

$$\mathbf{x} = \mathbf{s} + \mathbf{n}, \tag{4}$$

in which **x** and **n** are respectively: the desire signals matrix and noise with interference signals matrix.

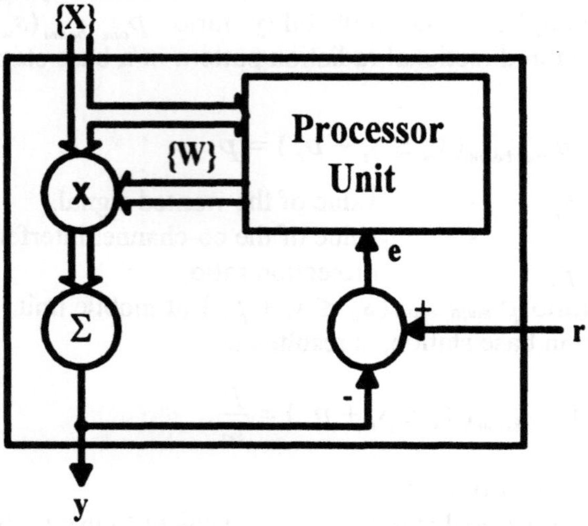

Figure 2. The adaptive processor structure

The difference between the output signal and the reference signal is the error signal:

$$e(j) = r(j) - \mathbf{w}(j)^T \cdot \mathbf{x}(j), \tag{5}$$

which should be minimaized by finding optimal values of the weights set, i.e.:

$$\mathbf{w}_{optimla} = \Phi^{-1}(x,x) \cdot \Phi(x,r), \tag{6}$$

where $\Phi(x,x)$ is a matrix of the crosscorelations and autocorrelations ratios of the received input signals, and $\Phi(x,r)$ is a vector of crosscorelations ratios between the input signals and the reference signals.

The knowledge of the direction of radio signal arrival, in short form so-called doa estimation, is very important to improve the efficiency of the optimal weights finding procedure, especially taking into account the real time requirements conditions.

4. DOA ESTIMATION PROCEDURES

From different methods which can be used for calculation of the doa, the two procedures: MUSIC and ESPRIT are extensively tested for application in adaptive processor scheme.

In the first step of the MUSIC method, the covariance matrix of the radio signals received by antenna array elements shoud be estimated by use the formal procedure:

$$\hat{\mathbf{R}}_{\mathbf{x}} = \frac{1}{J}\sum_{j=1}^{J}\left[\mathbf{x}(j)\cdot\mathbf{x}(j)^{H}\right] \tag{7}$$

in which the J is number of the radio signals samples. Next, by eigenvectors corresponds to the largest eigenvalues of the covariance matrix so the peaks in the direction of radio signal arrival can be obtained. The eigenvectors can be classified into two groups: one spanning the signal with noise subspace and one spanning a pure noise subspace. A matrix of the eigenvectors $\mathbf{e}$ is formed which span the pure noisespace. This matrix is orthogonal to the angles Θ from which the radio signal is arriving and it is possible to obtain high peaks in these direction Θ by solving the equation:

$$|d(\Theta)|^{2} = \sum_{i}\left|\mathbf{G}^{\mathbf{H}}(\Theta)\cdot\mathbf{e}_{i}\right|^{2} = 0, \tag{8}$$

in which $\mathbf{G}^{\mathbf{H}}(\ )$ is hermitian matrix of the directivity functions of the antenna array elements [8].

Due to the ESPRIT doa procedure the two Δ spaced parallel subarrays antenna elements are applied. The scheme to obtain the angles of radio signal arrival based on the eigenproblem analysis. First, the covariance matrix of the radio signals received by the two antenna subarrays should be estimated. Next, the angles Θ can be estimated by way:

$$\hat{\Theta} = \sin^{-1}\left\{\frac{c}{\omega\cdot\Delta}\cdot\arg\left(\hat{\Phi}\right)\right\}, \tag{9}$$

in which vector $\hat{\Phi}$ is estimated due to eigenvalues of the covariance matrix.

5. THEORETICAL INVESTIGATIONS

The directive properties of the adaptive antenna system with LMS optimization procedure improve by doa estimation were theoretically investigated. The modelled adaptive array was theoretically tested at 900 MHz common for GSM mobile communication system by use simulation system which consists of:

- PC computer for modelling of the mobile unit scenario,

- the signal processor for doa and LMS procedures realization.

Changes in the weight vector are made along the direction of the estimated gradient vector ∇ with scalar constant k_s to control rate of convergence and stability, i.e.:

$$\mathbf{w}(j+1) = \mathbf{w}(j) + k_s \cdot \nabla(j), \tag{10}$$

with

$$\nabla(j) = -2r(j) \cdot \mathbf{x}(j). \tag{10.1}$$

The obtained results of the influences of the adaptive process parameters on an attenuation of interference signals are exemplary presented in Figs. 3 and 4.

The influence of number N of the antenna array elements is shown in Fig. 3. The number of the antenna elements also depends on the require directivity properties of the antenna array and on the maksimum value of the mobile unit speed.

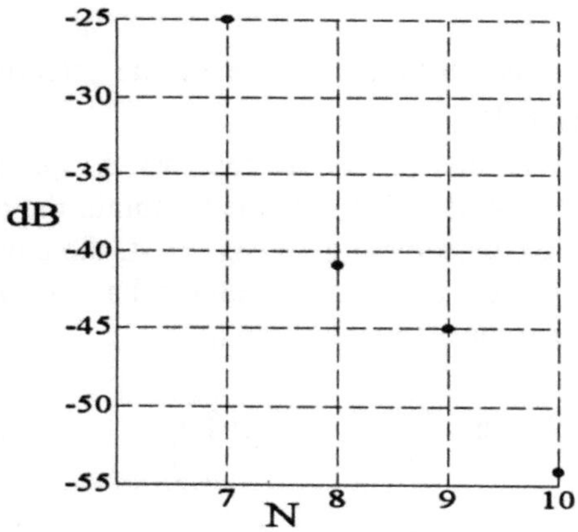

Figure 3. Interference signal attenuation vs. number of the antenna array elements

The influence of the number N_i of iteration steps on the interference signals attenuation is presented in Fig. 4. The optimal value of iteration steps depends on requirements to acceptable value of the interference signal attenuation and of a speed of the adaptive processor unit

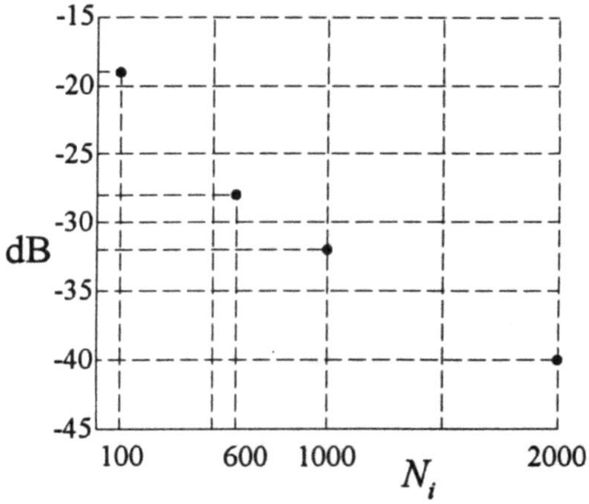

Figure 4. Interference signal attenuation vs. the number of iteration steps

Taking into account that the adaptive procedure is realized by iteration process, the maksimum time of single adaptation should be determined. The practically accepted maksimum time t_{max} depends on:

- the beamwidth β of the antenna array radiation pattern,
- the speed v of the mobile unit,
- and the distance R between the mobile unit and the base station.

The relation between the time t_{max} and the above mentioned parameters can be described as:

$$t_{max} = \frac{\pi \cdot \beta \cdot R}{360 \cdot v} \tag{11}$$

The limited values of the mobile unit speed obtained in described simulation experiments are presented in Table I. The obtained results show the requirements for quality of the adaptive signal processor, taking into account the real time scenario.

Table 1: Limited values of the mobile unit speed

$R(km)$	$v\left(\frac{km}{h}\right)$
0,1	1,8
0,5	9
4	36
6	72

At the and, the tested doa procedures can be characterized as:

- relative long calculation time is the main fault of the MUSIC algorithm,
- taking into account the inaccuracy of the antenna elements performance, the ESPRIT procedure is better that MUSIC.

6. CONCEPT OF THE ANTENNA SYSTEM

The microstrip antenna technology characterized by thin compact form is very attractive to application in modern radiocommunication systems, especially in mobile telecommunication. The proposed microstrip adaptive array consists patch rectangular radiating elements with circular polarization, to reduce the depolarization effect which is observed in mobile communication. To obtain the circular polarization the special two layer form of the microstrip radiating element corner fed should be applied. By stacking two patches in a double-layer structure, the antenna bandwith can be increased and dual-frequency properties can be obtained due to the two frequency subbands adequate to down and uplink operation [9].

The conceptional short form of the proposed microstrip adaptive array system is shown in Fig. 5.

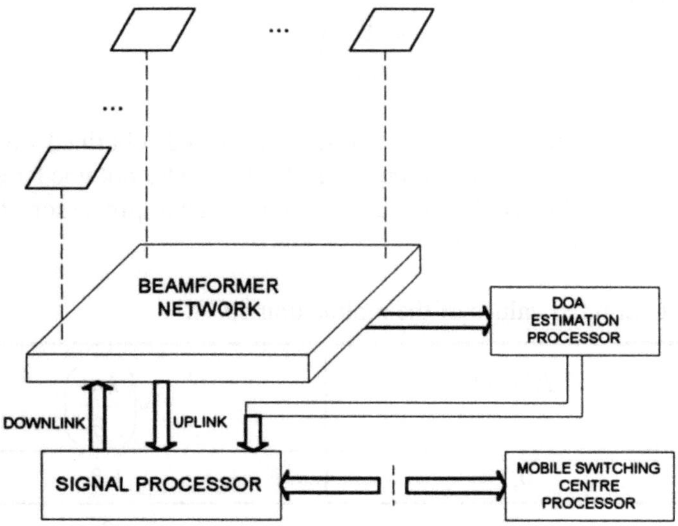

Figure 5. Concept of the microstrip adaptive array system

7. CONCLUSION

The obtained results show the usefulness of the theoretically tested adaptive antenna system for application in the base station of the mobile cellular telecommunication.

REFERENCES

[1] Macario R.C.V.: *Personal and Mobile Radio Systems*. IEE Telecomm. Series 25, Peter Peregrinius Ltd., 1991.

[2] Katulski R.J.: *Base Station Antenna System for Cellular Network*. Proc. of the Int. Wiss. Colloq., pp. 904-909, Germany-Ilmenau 1992.

[3] Katulski R.J.: *The EMC Antenna Aspect of Radio Systems Planning*. Proc. of the Int. Symp. on Electromagnetic Compatibility, pp. 604-607, Japan-Tokyo 1999.

[4] Hammuda H.: *Spectral Efficiency of Cellular Land Mobile Radio Systems*. Proc. of the 38th IEEE Vehicular Technology Conf., pp. 616-622, PA-Philadelphia 1988.

[5] Hubermark S.: *Smart Antennas - Intelligent Options*. Mobile Europe, pp. 51-58, March 1996.

[6] Dam H.: *Smart Antennas in the GSM System*. TSUNAMI TestBed Demonstration Seminar, no. 18/1, Aalborg University, Denmark, 1996.

[7] Compton T.: *Adaptive Antennas*. Prentice Hall, Inc., 1988.

[8] Katulski R.J.: *Algorithm MUSIC for a Cellular Network Application*, Proc. of the National Telecomm. Symp. KST-99, vol. D, pp. 117-123, Poland-Bydgoszcz 1999 (in Polish).

[9] Katulski R.J.: *Application of a Stacked Microstrip Antenna for Dual-Band Operation with Circular Polarization*. Proc. of the Int. Wiss. Colloq., pp. 73-77, Germany-Ilmenau 1995.

Robust Noise Reduction and Echo Cancellation

Kristian Kroschel, Martin Heckmann
Universitaet Karlsruhe, Institut fuer Nachrichtentechnik, Institut fuer Automation und Robotik
Kaiserstrasse 12, D-76128 Karlsruhe
kroschel@int.uni-karlsruhe.de

Key words: Robustness, noise reduction, echo cancellation

Abstract: Mobile phones used in cars require for safety reasons handset-free operation. This implies that a front-end is used which is able to reduce the ambient noise so that an acceptable intelligiblity is guaranteed for the far-end customer. Furthermore, the acoustic echo transmitted from the loudspeaker to the microphone has to be reduced so that it is not audible.
Another aspect has to be taken into account: the system has to be robust, i.e. it has to be operable in a wide range of the signal-to-noise ratio down to 0 dB or less. Furthermore the demand for a-priori knowledge should be as low as possible because acquisition of knowledge induces time delays unacceptable for real-time systems and contradicts a low-cost solution which is required for a mass product.
New developments in this paper concern the segmentation of the frequency band into subbands for all relevant components: noise reduction, echo cancellation and double-talk detection. Furthermore new concepts for echo reduction are presented.

1. INTRODUCTION

Handset-free communication from a car is of signi½cant interest: the main reason is safety since the driver should have all the time the ability to control the car manually which is not the case if he holds a mobile phone. Furthermore comfort is another aspect to use handset-free phones.

There are a couple of these phones on the market which are adapted to the special conditions in the car but still there is need for further research because the quality of the processed speech deserves further improvement and some of the commercially available systems degrade significantly in low signal-to-noise environments.

Two goals have to be met by handset-free phones: the first is to reduce the influence of the environmental noise so that the intelligibility of the transmitted speech is improved, the second is to reduce the speech echo from the far-end user which is generated on the path from the loudspeaker to the microphone.

In this paper ***robustness*** is the main issue of discussion. By robustness we understand that both operations, the noise reduction and the echo cancellation are as much as possible independent of the input signal-to-noise ratio, i.e they render an appropriate result and are stable even at low signal-to-noise ratios and require as few a-priori information as possible.

2. NOISE REDUCTION

There are many approaches for noise reduction. The most advanced ones are those based on signal estimation using Kalman filters [9]. This solution requires an appropriate model of speech and noise, respectively, which is in contradiction to our demand for a design which does not need much a-priori knowledge.

Since all model based systems require a lot of a-priori knowledge they all are discarded from our list of candidates for noise reduction. Thus we have to look for black-box solutions such as singular value decomposition, Karhunen Loève transform, vector quantization and others. They need much less a-priori information but the effort for mathematical computations is quite high since they need the calculation and inversion of correlation matrices. Furthermore, the quality of the processed speech might be degraded significantly since the parameters representing the uncorrupted speech are manipulated severely in those regions in which the parameters of the noise are strong. There are some ideas to overcome this problem which try to extract the lost information from the remaining parameters but this is an operation which requires a high effort.

A well-known solution for noise reduction, which is simple to implement, which is more or less independent of the input signal-to-noise ratio, and which does not depend too much on a-priori knowledge is based on Wiener filters [9] and known as ***spectral subtraction*** [14]:

$$H(e^{j\Omega}) = \frac{S_{SS}(e^{j\Omega})}{S_{RR}(e^{j\Omega})} = \frac{S_{RR}(e^{j\Omega}) - S_{NN}(e^{j\Omega})}{S_{RR}(e^{j\Omega})}$$

$$\approx \begin{cases} 1 - a \cdot \frac{S_{NN}(e^{j\Omega})}{S_{RR}(e^{j\Omega})} & 1 - a \cdot \frac{S_{NN}(e^{j\Omega})}{S_{RR}(e^{j\Omega})} \geq b \\ b & 1 - a \cdot \frac{S_{NN}(e^{j\Omega})}{S_{RR}(e^{j\Omega})} < b \end{cases} \quad (1)$$

with $S_{SS}(e^{j\Omega})$, $S_{RR}(e^{j\Omega})$ and $S_{NN}(e^{j\Omega})$ the power density spectra of the clean speech $s(k)$ and the speech corrupted by noise $r(k)$ and the noise $n(k)$, respectively. A block diagram of this system is given in Figure 1.

The a-priori knowledge required for this approach is the estimate of the noise power $S_{NN}(e^{j\Omega})$. This can be updated in speech pauses

$$\hat{S}_{NN}(e^{j\Omega}, i) = \alpha \cdot \hat{S}_{NN}(e^{j\Omega}, i-1) + (1-\alpha) \cdot |N(e^{j\Omega}, i)|^2 \quad (2)$$

with $N(e^{j\Omega}, i)$ the spectrum of the noise $n(k)$ in the speech pause with index i. The drawback of this approach is that highly instationary noise will be estimated quite unsatisfactorily. Furthermore, some time will pass by until the first speech pause is detected and an estimate of the noise is available. The requirement of a noise activity detector (VAD) can be counted under the negative aspects, too. There are two solutions without the need of a VAD: first, a microphone array might be used with the output of an estimate of the corrupted speech signal and of an estimate of the corrupting noise [7]. If this is assumed to be too expensive a simpler solution uses so-called minimum statistics [10], which means that an estimate of the noise power is permanently calculated as the minimum of the power in a specified time interval. The set-up time of this approach has proven to be very short in practical applications.

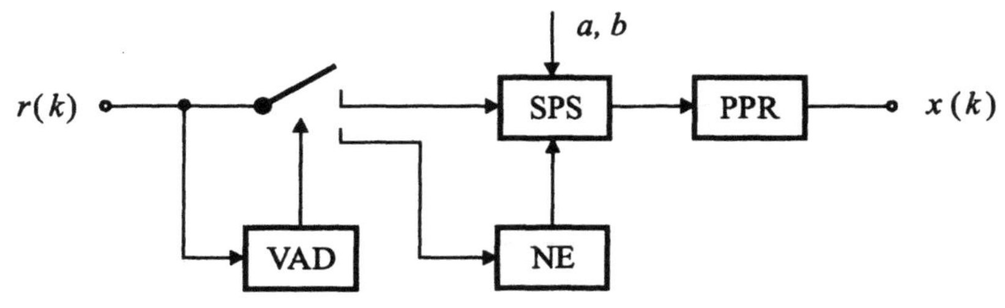

Figure 1: Noise reduction based on spectral subtraction (SPS). VAD: voice activity detector, NE: noise estimation, PPR: post processing.

The main drawback of spectral subtraction for noise reduction is that *musical tones* are produced which sound very unnatural and are very annoying. Spectral subtraction is parameterized by two values: the *overestimate* a with typical values $1 \leq a \leq 3$ and the *spectal floor* b with typical values $0.1 \leq b \leq 0.3$. With b set close to the upper limit the musical tones are made less audible but on the other hand the noise suppression effect is reduced so that intelligibility is affected. Two methods are suggested in literature to reduce the influence of musical tones: post-processing of data with a median filter of a typical lenght of $L = 3$ or $L = 5$ [2], an alternative is to exploit psycho acoustics [3], [6]. To avoid musical tones, those isolated spectral lines are set to the spectral floor which pass the auditory threshold and are not masked by the neighbouring spectral lines of the speech singnal.

Further modifications of spectral subtraction include a frequency dependent overestimate $a(\Omega)$ parameter [11] and the segmentation of the frequency range into specific bands.

3. ECHO CANCELLATION

The most satisfying solution for echo cancellation is the canceller based on the least mean square (LMS) algorithm [1], [15]

$$\mathbf{h}(k+1) = \mathbf{h}(k) + \mu(k) \cdot f(k) \cdot \mathbf{x}(k) \tag{3}$$

with $\mathbf{h}(k) = (h(1), h(2), \ldots, h(L))^T$ the vector of the impulse response $h(k), 1 \leq k \leq L$ of the echo path, $f(k) = \hat{y}(k) - y(k)$ the error of the echo $y(k)$ and its estimate $\hat{y}(k)$, $\mathbf{x}(k) = (x(k), x(k-1), \ldots, x(k-L+1))^T$ the input of the echo path and $0 < \mu(k) < 2/|\mathbf{x}(k)|^2$ the adaptation constant. A block diagram of this algorithm is given in Figure 2.

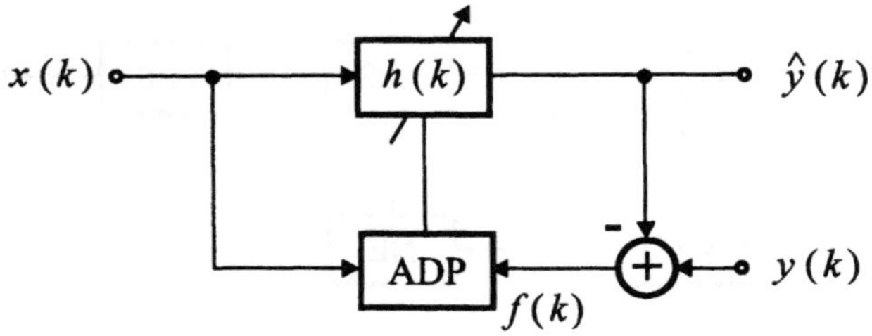

Figure 2: Least mean square estimation of impulse response $h(k)$. ADP: adaptive processing of parameters.

Since noise reduction is implemented in the frequency domain an algorithm for echo cancellation formulated in the frequency domain is preferred. In this case the signals are cut into blocks of lenght B. The result is the frequency least mean square (FLMS) algorithm [13]

$$\mathbf{H}(j+1) = \mathbf{H}(j) + \mu(j) \cdot \nabla(j) \tag{4}$$

with $\mathbf{H}$ the FFT of the augmented vector $\mathbf{h}(k)$, $\nabla(j) = g(\mathbf{X}^*(j) * \mathbf{F}(j))$ the gradient of the error $\mathbf{F}(j)$ which is the FFT of the augmented vector $\mathbf{f}(j) = (f(jB), f(jB-1), \ldots, f(jB-L+1))^T$ and $\mathbf{X}$ the FFT of the augmented vector $\mathbf{x}(k)$. Augmentation is required to avoid aliasing by applying overlad-add for the convolution.

These algorithms do not require a high computational load - e.g. no matrix inversion as the recursive least square (RLS) algorithm - and their adaptation to changes in the acoustical environment is fast enough for practical applications. The FLMS algorithm has the further advantage that the input values are more or less uncorrelated which improves the adaptation behaviour.

Due to changes of the acoustic transmission path between the loudspeaker and the microphone caused by movements of persons within the path or changes of the environment - opening and closing of windows and doors etc. - the echo canceller has to be adaptive. Speed and accuracy of the adaptation depend on the update parameter μ which is sensitive with respect to the signal-to-noise ratio and which is calculated for each frequency band separately [16].

To demonstrate the robustness of this approach, Figure 3 shows the behaviour of adaptaion in an adverse environment: with a low signal-to-noise ratio - SNR=0 dB - and a high influence of double talk - the power ratio of the echo and the local speech is 0 dB - the adaptaion of the frequency selective echo cancellation is superior to the conventional approach with an adaptation over the full spectral band. The compensation of the echo is expressed as echo return loss enhancement (ERLE) which fluctuates significantly due to the local speech. For comparison of the new and conventional approach, the system distance between the transfer function of the transmission path from the loudspeaker to the microphone and the adaptive echo canceller are given.

Noise is in this case the speech signal of the local speaker and the environmental noise. Adaptation of the echo canceller is executed only in periods in which the far-end speaker is active, i.e. an echo is generated. Therefore a voice activity detector is required in the loudspeaker path.

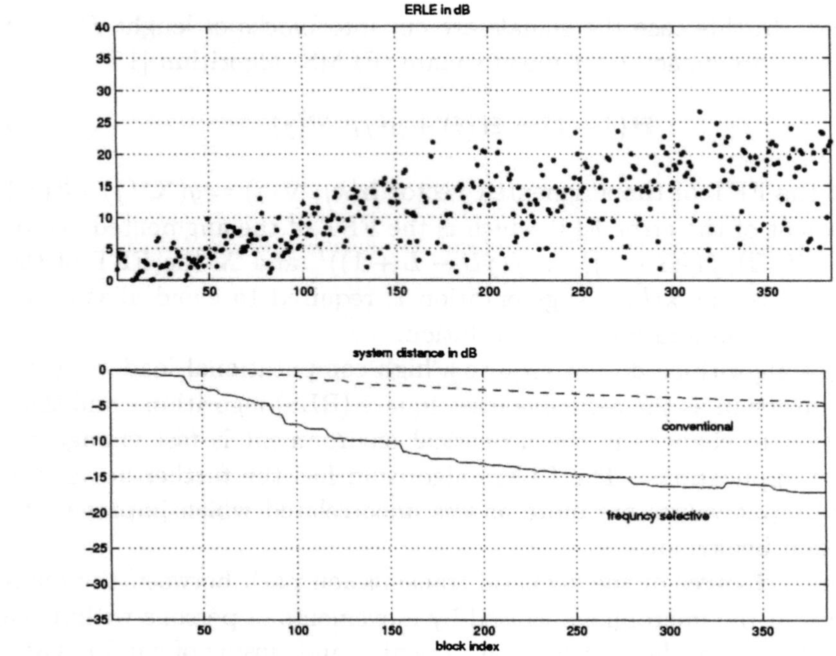

***Figure* 3:** Echo cancellation in an adverse environment: SNR=0 dB, power ratio of echo and local speech 0 dB.

The problem with the LMS and FLMS algorithms is that they rarely achieve the required attenuation of the echo specified by a given ERLE value. Therefore additional measures have to be taken. The one which will be discussed here further is the application of the so-called ***frequency selective gain control*** (FSGC) which allocates a gain value $H_i(e^{j\Omega})$ to the two transmission paths from the local speaker to the remote speaker and vice versa. The gain is frequency dependent and controlled by the magnitude of the speech spectra in the two transmission paths:

$$H_l(e^{j\Omega}) = \begin{cases} \beta_l \frac{|S_l(e^{j\Omega})|}{|S_f(e^{j\Omega})|} & \beta_l \frac{|S_l(e^{j\Omega})|}{|S_f(e^{j\Omega})|} < 1 \\ 1 & \text{elsewhere} \end{cases} \tag{5}$$

$$H_f(e^{j\Omega}) = \begin{cases} \beta_f \frac{|S_f(e^{j\Omega})|}{|S_n(e^{j\Omega})|} & \beta_l \frac{|S_f(e^{j\Omega})|}{|S_n(e^{j\Omega})|} < 1 \\ 1 & \text{elsewhere} \end{cases} \tag{6}$$

with l for the local speaker and f for the far-end speaker. According to Figure 3, $H_f(e^{j\Omega})$ is the transfer function of the far-end speech signal $x_f(k)$ to the output $y_f(k)$ of the local loudspeaker, e.g. $S_i(e^{j\Omega})$ denotes the spectrum of the transmitted speech which will be calculated by FFT. Comparing the magnitudes of the spectra of the local and the remote speaker the frequency axis is separated into bands. These bands can be as narrow as one spectral bin of the FFT. The width of these bands depends on the following aspects: the quality of the processed speech signal, the residual noise and the effort required for implementation.

With this approach double-talk communication is possible in contrast to conventional gain controls which cut off either one or the other transmission path. The desired echo attenuation is achieved by an appropriate reduction of the gain in each spectral band. This frequency dependent attenuation distorts the speech signal. But this distortion is almost not audible since those spectral components which are cut away from the spectrum of the local speaker are replaced by the corresponding spectral components of the far-end speaker. This is a very robust approach with the drawback that the speech signal is degraded for those listening to the dialogue of the local and the far-end speaker.

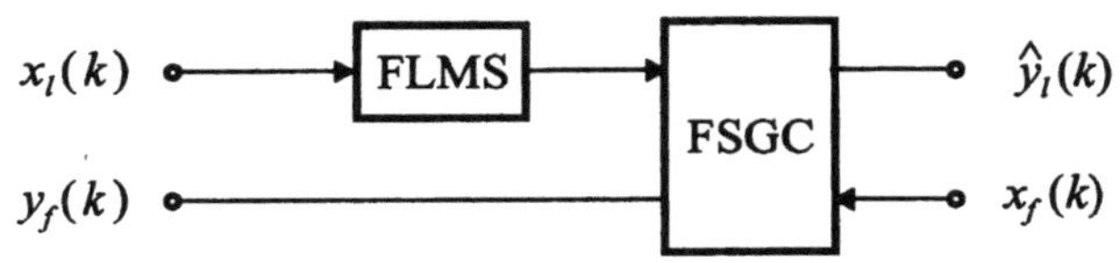

Figure 4: Combined echo cancellation. FLMS: frequency least mean square algorithm, FSGC: frequency selective gain control.

For the design of a robust echo canceller both approaches, echo compensation and frequency selective gain control, are combined. The idea is to exploit the absence of speech distortion of echo compensation for high signal-to-noise ratios and to use frequency selective gain control for low signal-to-noise ratios. Both components either can be arranged in parallel or serially. The parallel structure exploits the advantage that the speech signal might not pass the frequency selective gain control at high signal-to-noise ratios so that the speech signal is not distorted. The problem is to fuse the estimates of the echo gained from both systems. The fusion is based on a measure of quality of the estimated echo, and the overall estimate is the weighted sum of both individual estimates.

A more sophisticated approach subdivides the full frequency range into frequency bands and fuses the individual bands according to the measure of quality.

The cascade of the FLMS echo canceller and the frequency selective gain control FSGC given in Figure 4 is a simpler and robust solution: the FLMS component compensates the echo totally at high signal-to-noise ratios so that there is no need for further echo reduction by the subsequent FSGC. In this case $\hat{y}_l(k)$ is identical with the local speech. Thanks to the frequency selective adaptation of the FLMS even for low signal-to-noise ratios a significant contribution to echo reduction is gained. This is true even if the resulting ERLE gained by the FLMS is not sufficient. An example is given in the left part of Figure 5 which shows the echo reduction expressed in ERLE as a function of the data block index. It can be read from this result that some 150 data blocks are required for adaptation. The gained ERLE value of about 10 dB is not very satisfactory. The reason is the environmental noise with a signal-to-noise ratio of SNR=10 dB. The noise has been picked up in a car and is slightly instationary due to changes of speed.

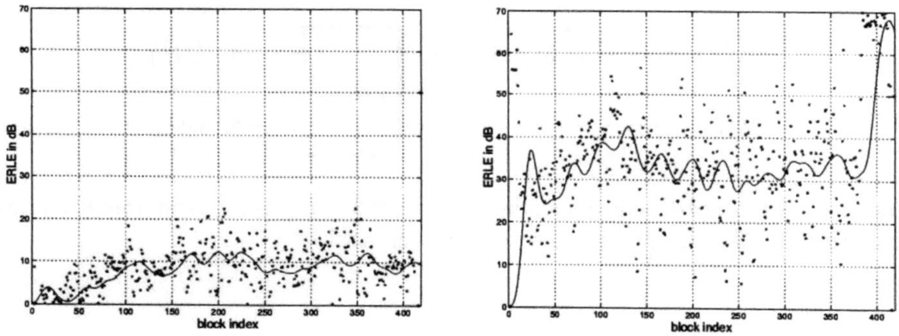

***Figure* 5**: **Echo cancellation: left by FLMS, right by the combination of FLMS and FSGC. Signal-to-noise ratio SNR=10 dB, power ratio of echo and local speech 0 dB. ERLE in dB as a function of the processed data block.**

The subsequent FSGC reduces the echo signifcantly as shown in the right part of Figure 5. The ERLE value is in the range of 30 to 40 dB which meets the requirements for practical application. In contrast to the FLMS algorithm the FSGC does not need any time for adaptation but renders almost from the start the required attenuation of the echo. The price for this operation is however an increase in distortion of both speech signals, the one transmitted to the far-end customer and the other one received locally.

4. COMBINING NOISE REDUCTION AND ECHO CANCELLATION

Integration of noise reduction and echo cancellation is a topic reviewed earlier in literature [5], and can further be combined with source coding [4], [8]. The latter is of no concern in this paper. Either both operations are executed in the time or the frequency domain. Since spectral subtraction and the frequency selective gain control require a realization in the frequency domain, echo compensation is realized on the basis of the FLMS algorithm to reduce implementational load.

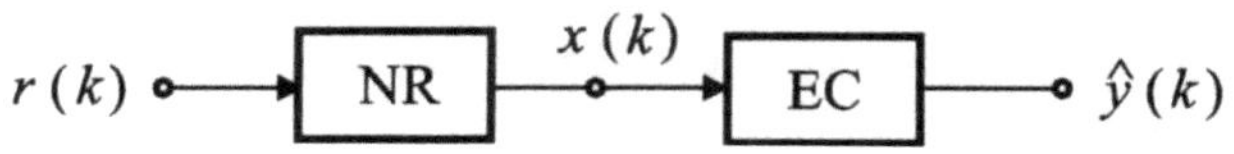

Figure 6: Combined noise reduction (NR) and echo cancellation (EC).

The question is how to integrate noise reduction and echo cancellation. A detailed investigation can be found in [12] where the question is answered whether noise reduction should precede echo cancellation or vice versa. Unfortunately a general answer cannot be given since it depends on the type of the applied noise reduction and echo cancellation method. Since noise reduction based on spectral subtraction is very robust we prefer the structure given in Figure 6 in which noise reduction is the first component followed by the echo canceller.

5. CONCLUSION

The system defined in this paper is an integration of well-known components. New developments are the segmentation of the frequency bands into subbands for all components: noise reduction, echo compensation together with the double-talk detector, and the frequency selective gain control. Furthermore, the combination of echo compensation with the frequency selective gain control and the interoperation of them are new ideas.

Tests with real-world signals have been executed for the LMS based, frequency selective echo canceller which showed that a significant gain in robustness in an adverse environment is gained. Informal listening tests have shown that the quality of the processed speech was appreciated by the listeners.

Further research will be cncentrated on the integration of the frequency selective gain control to improve the measure of echo cencellation and the spectral subtraction for noise reduction, respectively.

References

[1] Cowan, C.F.N.; Grant, P.M.: *Adaptive Filters.* Prentice Hall, Englewood Cliffs, 1985

[2] Haulik, T.: *Robuste Geraeuschreduktion fuer kompakte Mikrofonanordnungen.* PhD dissertation, Karlsruhe 1988

[3] Haulik, T.; Linhard, K.; Schroegmeier, P: *Residual Noise Suppression Using Psychoacoustic Criteria.* Proc. EUROSPEECH 97

[4] Kroschel, K.; Barros, J.: *Integrating Noise Suppression and Source Coding in Mobile Phones.* Proc. Electronische Sprechsignalverarbeitung, Goerlitz, September 1999, pp. 96-103

[5] Kroschel, K.; Ihle, M.: *Ein robustes System zum Freisprechen im Kraftfahrzeug.* Proc. Electronische Sprechsignalverarbeitung, Cottbus, August 1997, pp. 78-84

[6] Kroschel, K.; Kujaw, K.: *Speech Enhancement in a Noisy Environment Exploiting Phenomena of Hearing Physiology.* Proc. 43rd Int. Scientific Colloquium, TU Ilmenau, September 1998, pp. 439-444

[7] Kroschel, K.; Lange, K.: *Subband Array Processing for Speech Enhancement.* Proc. EUROSPEECH-93, Berlin, September 1993, pp. 621-624

[8] Kuropatwinski, M.; Lekschat, D.; Kroschel, K.; Czyzewski, A.: *Advanced Speech Signal Parameterization for Linear Predictive Coder Based Speech Enhancement,* ISSEM'99, Gdansk, September 1999, pp. 43-51

[9] Kroschel, K.: *Statistische Nachrichententheorie.* 3rd ed, Springer, Berlin etc., 1996

[10] Martin, R.: *An efficient algorithm to estimate the instantanous SNR of speech signals.* Proc. EUROSPEECH-93, Berlin 1993, pp. 1093-1096

[11] Mekhaiel, M.: *Untersuchung und Vergleich nichtlinearer Geraeuschreduktionsverfahren fuer robuste Spracherkennung.* Diploma Thesis, Institut fuer Nachrichtentechnik, Universitaet Karlsruhe 1994

[12] Scalart, P.; Benamar, A.: *A system for speech enhancement in the context of hands-free radiotelephony with combined noise reduction and acoustic echo cancellation.* Speech Communication 20 (1992), pp. 203-214

[13] Soo, J.S.; Pang, K.K.: *Multidelay block frequency domain adaptive filter.* IEEE Trans. Acoustics, Speech and Signal Processing 38(2) (1990), pp. 373-376

[14] Vary, P.: *On the Enhancement of Noisy Speech.* Proc. EUSIPCO-83, Erlangen 1983, pp. 327

[15] Widrow, B.; Hoff Jr., M.E.: *Adaptive switching circuits.* IRE WESCON Conv. Rec. 1960,vol 4, pp. 96-104

[16] Heckmann, M.; Vogel, J; Kroschel, K.: *Frequency selective step-size control for acoustic echo cancellation.* Accepted for publication in Proc. EUSIPCO 2000, Sept. 4-8, 2000, Tampere, Finland

Estimation of the Channel Impulse Response for GSM System

Jacek Stefański
Technical University of Gdansk, Department of Radiocommunication
e-mail: jstef@eti.pg.gda.pl

Keywords: GSM, MLSE, training sequence.

Abstract The aim of this article is to present an improved training sequence for estimating the non-stationary channel impulse response used in GSM.Many components of the received signal in digital communication systems arrive to the receiver antenna with different delays. These delayed signal components can cause intersymbol interference, increase BER (**B**it **E**rror **R**ate) and hence degrade quality of the received source information. Therefore, in order to increase the quality of received signal, every transmitted data packet contains a training sequence. However, the training sequence recommended by GSM is not the optimal method of estimating the channel impulse response. This can be clearly portrayed in a simulation performance.

1. INTRODUCTION

In digital cellular mobile communication systems, such as GSM (Global System for Mobile Communications), intersymbol interference which occurs due to a time-variant multipath fading must be neutralized by the application of adaptive equalizers.

A MLSE (Maximum Likelihood Sequence Estimator) represents the optimal receiver structure [1]. The MLSE receiver consists of matched filter and a Viterbi processor. The received signal is sampled and each sample is filtered through a matched filter whose parameters are approximated by training sequence. Viterbi processor equalizers estimate transmitting symbol sequence by using the Viterbi algorithm.

The number of matched filter taps depends on the maximal echo delays which in turn determine the number of states in the Viterbi processor (the number of matched filter taps for GSM is five).

2. GMSK MODULATION

A GMSK (Gaussian Minimum Shift Keying) [2] modulated signal can be represented as

$$s(t) = e^{j\theta(t)}, \theta(t) = \theta_0 + \sum_i b_i \phi(t - iT_b) \tag{1}$$

where θ_0 is an initial phase, T_b is the bit period and $b_i \in \{1, -1\}$ are the differentially encoded data bits. In terms of the raw data bits $a_i \in \{0, 1\}$, $b_i = 1 - 2(a_i \oplus a_{i-1})$, where $\oplus$ denotes modulo 2 addition [3]. The phase pulse-shaping function $\phi(t)$ is given by

$$\phi(t) = K \cdot \int_{-1,5T_b}^{t} y(\tau) d\tau \tag{2}$$

where: $y(t) = \frac{A}{\sqrt{2\pi}} \left\{ Q\left[\frac{2\pi BT_b}{\sqrt{\ln 2}} \left(\frac{t}{T_b} - \frac{1}{2} \right) \right] - Q\left[\frac{2\pi BT_b}{\sqrt{\ln 2}} \left(\frac{t}{T_b} + \frac{1}{2} \right) \right] \right\}$,

$$Q(x) = \frac{1}{\sqrt{2\pi}} \int_x^{\infty} e^{-\frac{t^2}{2}} dt,$$

$BT_b = 0.3$ (for GSM),

A, K - constans.

To implement the Viterbi algorithm, it is imperative to determine a finite number of states for the GMSK modulated signal. This can be accomplished by modifying the phase pulse-shaping function $\phi(t)$ from fig. 1. $\phi(t)$ is approximated as [4].

$$\hat{\phi}(t) = \begin{cases} 0 & for \quad t < -T_b \\ \phi(t) + \phi(-1{,}5T_b) - \phi(-1{,}5T_b - t) & for \quad -T_b < t < -\dfrac{T_b}{2} \\ \phi(t) & for \quad -\dfrac{T_b}{2} < t < \dfrac{T_b}{2} \\ \phi(t) + \phi(1{,}5T_b) - \phi(2{,}5T_b - t) & for \quad \dfrac{T_b}{2} < t < T_b \\ \dfrac{\pi}{2} & for \quad t > T_b \end{cases} \tag{3}$$

Following the above approximation there are only eight possible transmitted complex signals during the symbol period $[(n-0{,}5)T_b, (n+0{,}5)T_b]$ (one set of eight complex signals for even n and another set of eight for odd n), where n is an arbitrary integer.

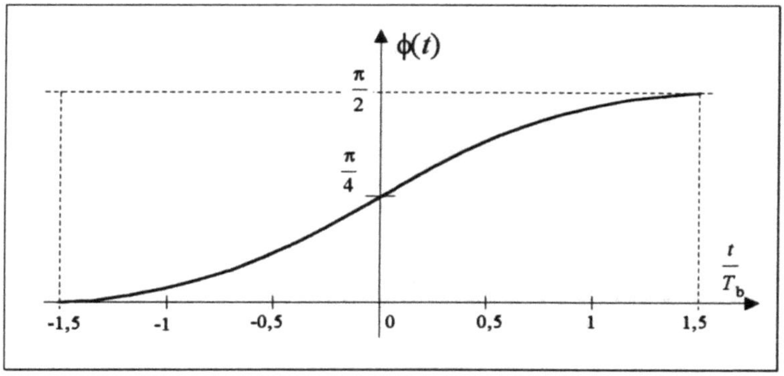

Figure 1. Phase pulse-shaping function.

The state transition trellis diagram for the GSM modulated signals is shown in fig. 2.

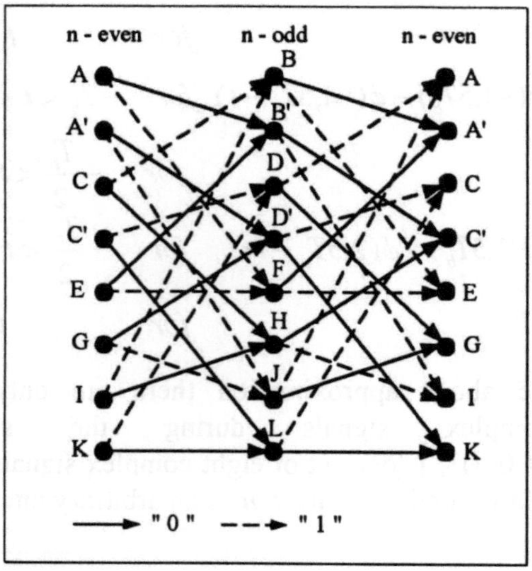

Figure 2. Trellis diagram of the GMSK modulated signal.

It should be noted that the trellis structure remains the same no matter what direction the transition comes from; from even to odd states or from odd to even states.

3. GSM BURST STRUCTURE AND CHANNEL ESTIMATION

The standard GSM burst structure is shown in fig. 3.

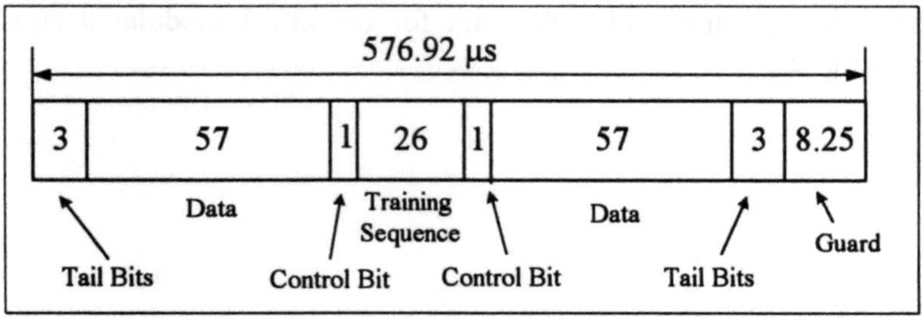

Figure 3. GSM burst structure.

The two 57 bits data fields are separated by two control flags and a 26 bit training sequence. This training sequence is used within GSM receiver for a precision synchronization and an estimation of channel impulse response.

The estimate may be obtained for instance by correlating the received training sequence with a local copy held at the receiver.

Altogether eight such sequences have been defined within the GSM recommendations [3]. These sequences have been selected based on their good autocorrelation properties and their low cross-correlation properties between one another. Each sequence is composed of three distinct sub-sequences X, Y and Z, with sub-sequences X and Z being used twice within the entire sequence (fig. 4 [5]).

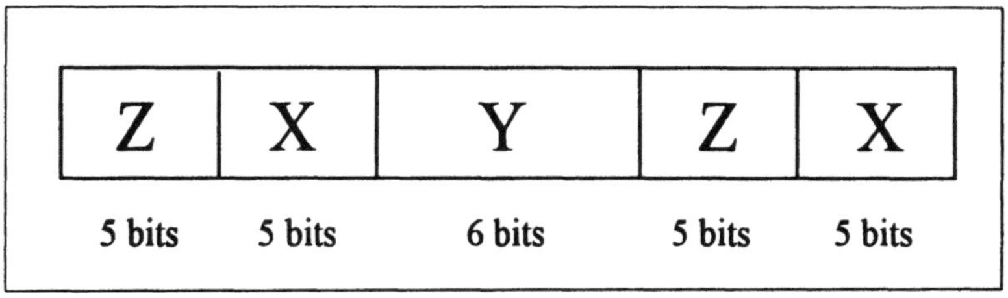

Figure 4. GSM training sequence structure.

The middle 16 bits in the entire 26 bits GSM training sequence allow the receiver to estimate the channel impulse response using five complex taps. It is crucial to note that all sequences share the central autocorrelation function peak surrounded by five „0" on each side. All possible sequences have been thoroughly analyzed. The autocorrelation results are portrayed in Fig. 5. Figure 5a presents the autocorrelation function of one of eight training sequences recommended by GSM, whereas figure 5b presents the autocorrelation function of the improved training sequence calculated between the central 16 bits.

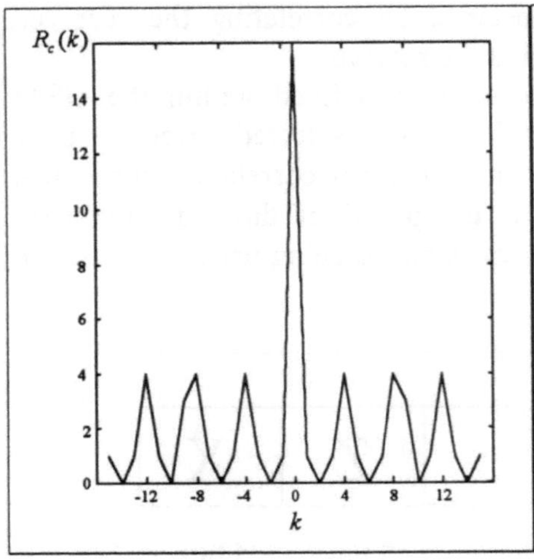

Figure 5a. Autocorrelation of one of the recommended training sequence used in GSM system.

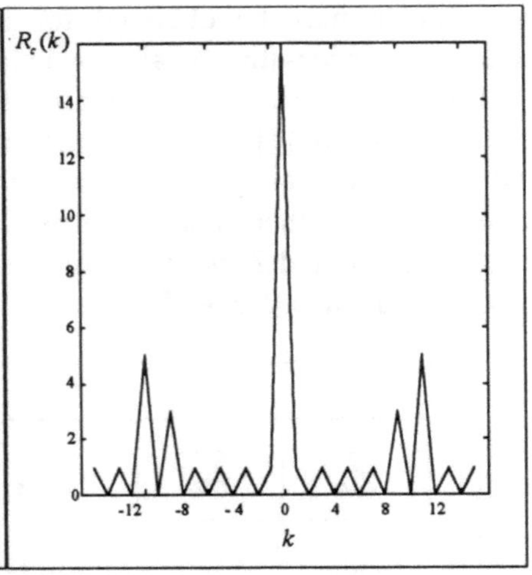

Figure 5b. Autocorrelation of the improved training sequence.

By comparing the two pictures, it can be observed that the improved autocorrelation function of training sequence is clearly better than that recommended by GSM. This can also be shown by simulation performance.

4. GSM RADIO LINK SIMULATION

The GSM simulation tool was designed according to ETSI (European Telecommunication Standard Institute) specifications [3]. The simulator has a flexible, modular structure in which each main GSM system element forms a basic simulation block. The simulator consists of: bits generator, channel encoder, interleaver, modulator GMSK, radio channel (with 3 propagation profiles TUx - Typical Urban, HTx - Hilly Terrain and RAx - Rural Area, where x is the vehicle speed [km/h] [6]), adaptive whitened matched filter, detector MLSE (with 8-state Viterbi algorithm), deinterleaver and channel decoder. Using the GSM simulator a number of packets containing the training sequence recommended by GSM and the improved version of training sequence were sent through a three channel models. The estimated BER performances for the TU50, HT100 and RA250 channel models are shown in fig. 6 ÷ fig. 8.

In channels with a low dispersion of less than $5\mu s$ (TU and RA profiles) a good performance for estimating the channel impulse response can be achieved by using both training sequences. However, the results are slightly better (about 0.1 ÷ 0.3 dB) by using the new training sequence.

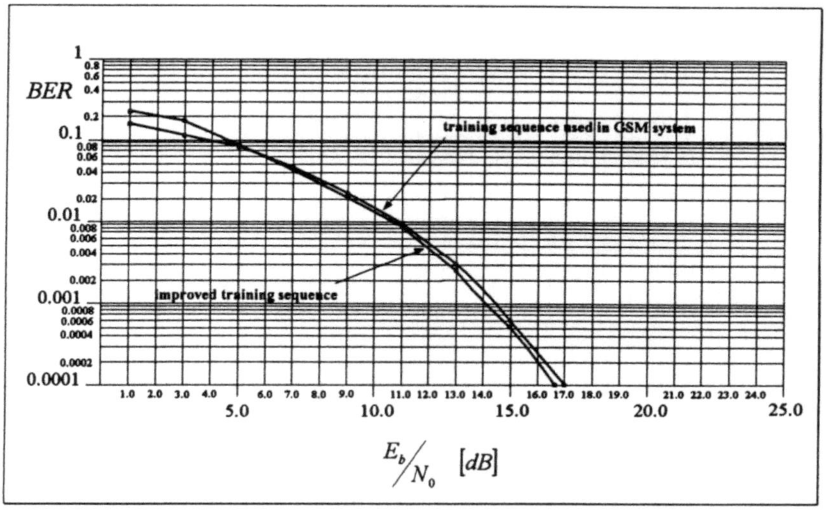

Figure 6. Simulation results under TU50.

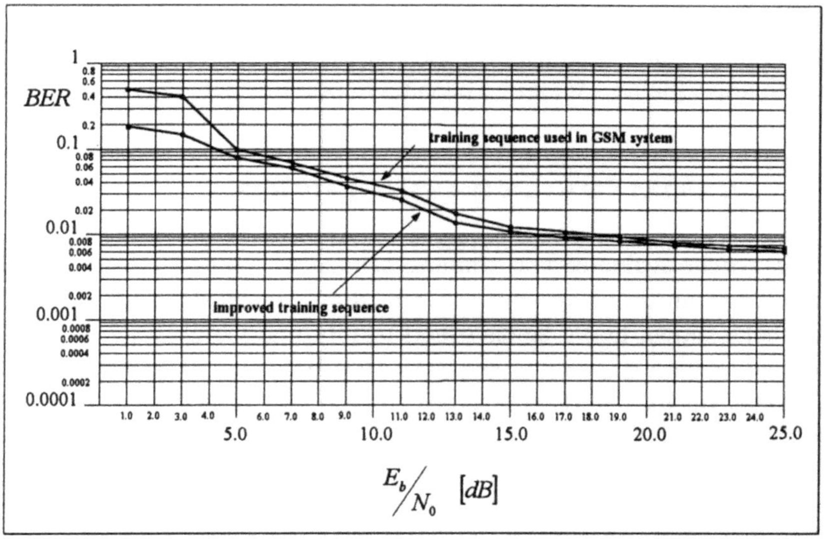

Figure 7. Simulation results under HT100.

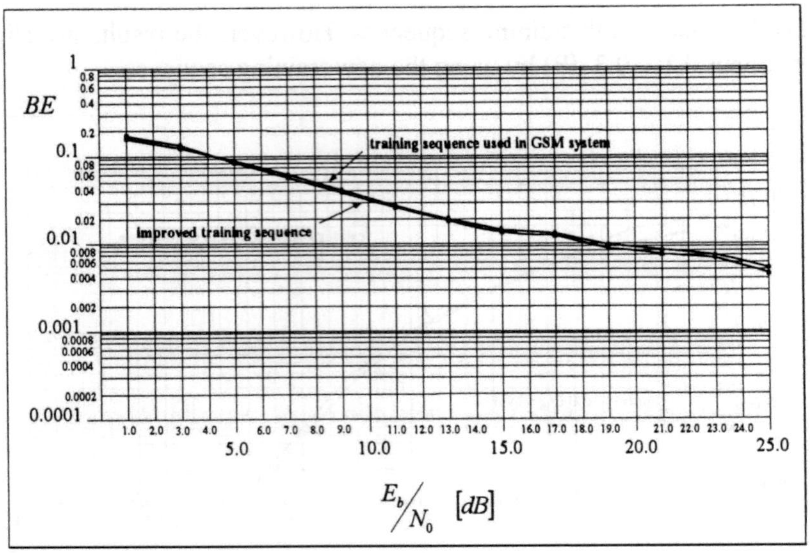

Figure 8. Simulation results under RA250.

In channels with a higher dispersion (HT profile), however, the results are clearly better by using the new training sequence (about 1dB). This can by explained by the fact that the shape of the autocorrelation function of a new training sequence resembles the shape of the autocorrelation function of white noise (better estimation channel impulse response) [7].

5. SUMMARY AND CONCLUSIONS

The article evaluates two training sequences applied in digital cellular mobile communication systems for estimating the channel impulse response. It has been shown that a better sound quality may be obtained by using an improved version of training sequence than by using the training sequence recommended by GSM. Therefore, it can be argued that the training sequence used in GSM system has been not selected based on maximizing the quality of received signal through minimizing BER but rather on ensuring the maximum synchronization in the system.

REFERENCES

[1] Forney G.: *Maximum-Likelihood Sequence Estimation of Digital Sequences in the Presence of Intersymbol Interference*. IEEE Transactions on Information Theory, vol. IT-18, no. 3, pp. 363-378, May 1972.

[2] Murota K., Kenkichi H.: *GMSK Modulation for Digital Mobile Radio Telephony*. IEEE Transactions on Communications, vol. COM-29, no. 7, July 1981.

[3] *GSM Technical Specifications*. ETSI, Sophia Antipolis, 1992.

[4] Chen J. T., Paulraj A., Reddy U.: *Multichannel Maximum-Likelihood Sequence Estimation (MLSE) Equalizer for GSM Using a Parametric Channel Model*. IEEE Transactions on Communications, vol. COM-47, no. 1, pp. 53-63, January 1999.

[5] Joyce R. M., Ibbetson L. J., Lopes L. B.: *Prediction of GSM Performance Using Measured Propagation Data*. Proc. 46th Vehic. Technol. Conf., pp. 2246-2250, 1996.

[6] *COST 207. Final Report, Digital Land Mobile Radio Communications*. Commision of the European Communities, Luxemburg, 1989.

[7] Nahi N. E.: *Estimation Theory and Applications*. John Wiley & Sons, 1969.

Keyword Index

AAL2, 53, 133
ABR, 17, 43, 133
adaptive antenna system, 239
adaptive processor, 239
AMICA, 1
amplifier, 227
ATM, 17, 43, 53
BCH, 79
broadband WATM, 169
BTS, 101
CAC, 17, 43, 67, 133
CBR, 17, 43, 133
CDMA, 17, 67, 89, 133, 169
cellular systems, 67, 79, 89, 101
channel coding, 79
  model, 227
  reservation, 43, 133
CMR, 101
COST 257 project, 169
CRC, 79
CSMA/CA, 29, 157
CTS, 157
echo cancellation, 249
error protection, 79
ETSI, 89
EY-NPMA, 29
FA, 199
FDD, 89
FDMA, 89
flow control, 169
GMSK, 259
GPRS, 147
GSM, 79, 89, 227, 259
HA, 199
handover, 1, 213
  latency, 213
hidden stations, 157
HIPERLAN, 29
HLD(HLR), 101
H.261,147
IEEE 802.11, 29, 14, 157
IGMP, 213
IMT2000, 17, 89
interference range, 89
IP, 53
  mobility, 199, 213
  multicast, 213
LA, 101
LEO, 17
location management, 101

MAC protocols, 29, 43, 133, 169
MEDIAN, 133
membership management, 213
microcell, 123
microstrip technology, 239
MLSE, 259
mobile communication, 199
mobile IPv4, IPv6, 199
mobile systems, 227, 239
mobility, 1, 101, 123
modelling, 123
MSA, 213
MSC, 101
multimedia applications, 169
multimedia traffic, 133, 147
narrowband networks, 169
new services, 169
network architecture, 169
noncooperative scheduling, 29
nonlinearity, 227
performance analysis, 133, 157, 227,
picocells, 89
predistortion, 227
PRMA, 17, 133
QoS, 17, 53, 133, 147, 169
RAN, 53
real time communication, 199
random access chanel, 43
random token, 29
resource allocation, 43, 67
  utilization, 17
route optimization, 199
RTCA, 29
RTS, 157
robustness, 249
satellite communications, 17, 53
scheduling policies, 29
SNR, 17
speech coding, 79
speech decoding, 79
speech processing, 79, 249
TDD, 43, 133
TDMA, 43, 89, 133, 169
TDMA/TDD, 43, 133
TINA,101
training sequence, 259
traffic control, 17
UBR, 17, 43, 133
UMTS, 53, 101, 123, 227
USRAN, 53
UTRAN, 53
VBR (rtVBR, nrtVBR), 17, 43, 133
video codec, 147
Viterbi processor, 259
VLD(VLR), 101
WAP, 101
WATM, 43, 133
WCDMA, 89
wireless LANs (WLANs), 29, 157
  internet, 1
  networks, 1, 169
3G, 53, 123, 147, 227

MIX
Papier aus verantwortungsvollen Quellen
Paper from responsible sources
FSC® C105338

If you have any concerns about our products,
you can contact us on
ProductSafety@springernature.com

In case Publisher is established outside the EU,
the EU authorized representative is:
Springer Nature Customer Service Center GmbH
Europaplatz 3, 69115 Heidelberg, Germany

Printed by Libri Plureos GmbH
in Hamburg, Germany